Computer-Aided Design

Manufacturing Technology Research

Mechanical Engineering Theory and Applications

Dimitrios Tzetzis
Panagiotis Kyratsis

Computer-Aided Design

Advances in Research and Applications

nova
science publishers
New York

Library of Congress Cataloging-in-Publication Data

ISBN: 979-8-88697-609-0

Published by Nova Science Publishers, Inc. † New York

Contents

Chapter 1

Employing Virtual Reality for the Design of Human-Robot Collaborative Workstations

Dimitris Nathanael* and Loizos Psarakis

School of Mechanical Engineering, National Technical University of Athens, Zografou, Athens, Greece

Abstract

In the last years, with the continuous advances in technology there has been a great increase in the use of robots in industrial manufacturing settings. New hybrid workstations are established where the human collaborates with one or even more robots sharing the same workspace along with common plans and goals. Designing these hybrid assembly work cells for heterogeneous agents presents significant challenges such as human safety and team effectiveness. The use of computer-aided design technologies, particularly virtual reality (VR), bears novel possibilities for human–robot collaboration (HRC) in both the design and training phases, as they are cost effective and time saving, while they offer easy prototyping along with controlled and safe settings. The effectiveness of VR simulations, apart from immersion characteristics, is crucially dependent on the resemblance of user experience between virtual and real environments. In the field of HRC, several researchers have employed VR simulation as a study tool utilizing the various advantages it offers. The chapter first reviews recent developments and pending issues in the field; then a case study developed by the authors is presented, where an industrial human–robot collaborative sorting task was designed, implemented, and tested by volunteer participants, in a VR simulation environment.

Keywords: human–robot collaboration, VR simulation, work cell design

Introduction

Nowadays, robotics is at the threshold of a new Industrial Revolution. As artificial intelligence radically grows and cloud technologies, such as the Internet of Things and the 5G, flourish, robots start to become a part of our everyday life. According to the International Federation of Robotics, there has been a great increase in the production of robots in the last decade (International Federation of Robotics, 2021a,b). Particularly, in industrial settings the annual robot installations were increased by 13% on average each year since 2014, while in the field of service robots there has been a great increase in 2019 of 32% in professional service robots

* Corresponding Author Email: dnathan@central.ntua.gr

and 40% in service robots for domestic and household tasks. The industry is entering the fourth industrial revolution (Industry 4.0), where smart factories integrate new technologies and robots start to communicate with each other as well as with human coworkers, resulting in the improvement of product manufacturing and distribution. Even though there is still widespread concern over the potential jobs losses (Maurtua et al., 2017; Müller-Abdelrazeq et al., 2019; Weiss et al., 2011; Takayama et al., 2008), the researchers and the industry show increasing interest in establishing new hybrid human–robot collaboration (HRC) workstations. Undoubtedly, the incorporation of human cognitive skills, like contextual flexibility, problem-solving, and dexterity, with the robot's advantages, like high speed, repeatability, power, and precision, promises significant benefits such as increased productivity and robustness along with better work conditions (Wang et al., 2019a; Koppenborg et al., 2017; Wang et al., 2020). Still, designing and testing these new hybrid assembly work cells presents major concerns such as maintaining human safety and achieving team communication between heterogeneous agents. Generally, "a collaboration occurs when a group of autonomous stakeholders of a problem domain engage in an interactive process, using shared rules, norms, and structures, to act or decide on issues related to that domain" (Wood and Gray, 1991). In HRC the human collaborates with one or even more robots in a mutual workspace, with mutual plans, goals, and rules while they seamlessly understand each other (Psarakis et al., 2022).

Research and training in the field of HRC remains a risky process, especially for the safety of human operators, as well as a tedious task since it requires skillful robot programming and large-scale manual integration. To alleviate some of these challenges, immersive virtual reality (VR) technologies have been employed to simulate collaborative environments with considerable success in many work domains. Actually, in any work situation, where either the cost or the possible negative consequences of testing design solutions to the real task environment are considerable, VR systems may provide a valuable alternative.

Experimenting in a virtual environment is beneficial for HRC in both the design and training phases. Specifically, designing a collaborative task in VR (i) is cost effective as expensive real robots are not necessary, (ii) is time saving as the production is not ceased for repeated testing while the changes and redesigns are easily applied, (iii) provokes novel perspectives by providing the visualization of new angles, and finally (iv) allows easy transportability of the technical systems used which are constantly becoming even more inexpensive. On the other hand, training the human operators in VR systems before their collaboration with the real robots principally offers controlled and safe settings while supporting immersive naturalistic interactions with virtual objects and precise and accurate measures (such as the kinematics of the movements made, collisions, and task completion times).

Effectiveness of VR in Simulating Real Work

Even though VR technologies are flourishing today, their introduction dates back to the middle of the 20th century, when the first head mounted display (HMD) devices were developed. In particular, in 1956 Morton Heilig invented the first 3D immersive simulator called Sensorama (Heilig, 1962), while a few years later in 1965 Ivan Sutherland presented the ultimate display, describing it as a device that with the right programming the virtual world would not be distinguished from the real (Sutherland, 1965). Sutherland visioned various emergent technologies to endorse his devices such as trackpads, dynamic perspective rendering, voice recognition, haptics, and eye-gaze tracking (Sutherland, 1965, 1968). Since then, VR systems have greatly evolved to the present high-quality and low-cost portable devices such as the Oculus Rift™ and Quest™, the HTC Vive™, and the PlayStation VR™.

As people tend to believe, the effectiveness and the applicability of a VE are strongly related to the necessary physical fidelity to mimic the resolution of the physical world (Gupta et al., 2008). Specifically, it is generally accepted that the fundamental parameters of successful virtualization are presence and immersion. Presence is a phenomenological term and is defined as the sense of being in a virtual environment rather than at the place in which the participant's body is actually located. On the contrary, immersion is a technical term and can be defined as the replacement of as many real-world sensations as possible, by sensations corresponding to the virtual environment. Immersion is by essence related to the multimodal nature of the perceptual senses, and to the interactive aspects of a VR experience. The term immersion, thus, stands for what the technology delivers from an objective point of view. The more a system's displays and tracking preserves fidelity in relation to their equivalent real-world sensory modalities, the more it is immersive (Slater and Wilbur, 1997; Sanchez-Vives and M. Slater, 2005).

A challenge for VR effectiveness and applicability has to do with the side effects experienced during VR simulation, namely general discomfort, fatigue, and dizziness. A few theories seek to explain the VR sickness phenomenon such as the sensory conflict theory (Reason and Brand, 1975; Oman, 1982) or the postural instability theory (Riccio and Stoffregen, 1991). Besides these theories, there are also several technical aspects of VR that can induce sickness such as the refresh rate, the response time, the projected images, and the resolution of the HMDs. Especially, in the first modern VR technologies the aforementioned technical characteristics of the systems were not adequate and resulted in frequent dizziness formulation to the subjects. Therefore, the designers of the virtual environment had to deal with this disadvantage. Typical solutions involved minimizing the movement of the subjects or even preferring a sitting interaction position to a standing one or removing fast-changing and flashy content from the background. Nevertheless, with continuous advances in computing and visualization technologies, these issues are being steadily reduced.

However, the belief that successful VR simulations depend solely on the quality of immersion or even presence has been questioned (Gopher, 2012). One should not forget that some of the seminal VR simulators of work situations such as the MIST VR surgical simulator (Gallagher et al., 1999) have been highly successful even though they are judged as of very low immersion by today's standards. Technical refinements and advances will always be welcome, but they alone cannot guarantee successful simulation. In the context of training, Gopher (2012) proposes that the value of a VR system should be judged by its ability to provide a resembling experience, by the provision of facilitation and guidance to the acquisition of the designated skill and by the transfer from VR training to performance in the real world. According to this view, relevance, facilitation, and transferability are crucial evaluation criteria for a VR simulator. Therefore, in order to become effective, VR should be oriented toward the establishment of what it is that is being transferred from the virtual to the real environment (Rose et al., 2000).

Following the above, the effectiveness of VR in simulating real-world work situations is crucially dependent on the resemblance of user experience between virtual and real environments. Resemblance of experience between VR and real environments seems to vary depending on the type of task simulated. For example, a considerable body of research suggests that for spatial skills and procedural tasks VR provides an experience that resembles quite well with the real environment (Regian, 1997; Brooks, 1999; Waller et al., 1998; Aurich et al., 2009). On the contrary, a frequent criticism is that VR is rarely adequate for tasks that are characterized by a significant perceptual and/or motor component (Bergamasco et al., 2012), such as in micro-surgery or in machining. In such tasks, resemblance of the user experience in VR compared to the real environment is quite low. For this reason, various ways have been used to enhance VR experience resemblance both in cognitive and sensorimotor tasks such as augmented reality (AR) (Ong et al., 2008) or by designing virtual cognitive and sensory enhancements (Nathanael et al., 2016).

Several studies have tackled the issue from a task analysis point of view. Dankelman et al. (2003) and Wentink et al. (2003), in a surgical context, have used Jens Rassmussen's (1983) classification of human behavior at the skill–rule–knowledge levels to discuss the effectiveness of VR simulators. In the CNC domain, Lin et al. (2002) used goal decomposition, that is, a variation of hierarchical task analysis to represent task execution scenarios and associated simulator objectives. Bergamasco et al. (2012) have reported significant progress in VR training by concentrating on detailed and exhaustive analysis of skills prior to VR simulator implementation. Bardy et al. (2012) propose a pragmatic decomposition of sensorimotor skills into thirteen quasi-independent functional sub-skills. This decomposition is taken as a starting point for focusing VR simulator training on specific elements. The above work has demonstrated that a preliminary analysis of a task in terms of cognitive, perceptual, and motor abilities needed for its performance may significantly affect experience resemblance and skill transfer between virtual and real work situations.

VR Simulators for HRC

Over the last few years, several researchers have employed VR simulation as a study tool for HRC, utilizing the various advantages it presents (Nathanael et al., 2016a; de Giorgio et al., 2017; Matsas et al., 2018; Rückert et al., 2018; Dimitrokalli et al., 2020; de Freitas et al., 2022; Malik et al., 2020; Psarakis et al., 2022). The main research issues on HRC concern human safety, human–robot communication, and overall team performance and efficiency. Specifically, regarding the safety of the humans collaborating with the robots, researchers mainly focus on reducing contact forces and payloads, on strategies for collision avoidance, on the integration of safety zones and stop functions, on various ergonomic issues, and on speed and trajectory modifications (Lasota et al., 2017; Robla-Gómez et al., 2017; Michalos et al., 2015; Kim et al., 2019). Furthermore, the study on human–robot communication can be broadly divided into: (i) the way the collaborating agent communicates information, such as gaze behaviors, gestures, visual cues, intent expressions, visual boundaries of the working envelope, and alerts (Huang and Mutlu, 2016; Baraka et al., 2016; Matsas et al., 2018; Liu and Wang, 2018), and (ii) the way the collaborator receives information such as the use of sensors' data, interaction history models, control algorithms, cost/reward based frameworks, and legible motions (de Gea Fernández et al., 2017; Liu et al., 2018; Nikolaidis et al., 2015b). Finally, research works on overall team performance and efficiency principally focus on task sequence and allocation, team fluency, optimized movement paths and trajectories, collaborating agent behaviors, work cells' layout, minimizing the makespan, and applying the optimal robot speed (Hoffman, 2019; Psarakis et al., 2022; Faccio et al., 2020).

Designing a virtual environment for HRC investigation includes the human avatar, the robot simulator, and the various 3D objects that complete the environment. Simulating a robot bears greater complexity as it should resemble a real model that is able to function properly. Robot simulators are based on computer-aided design (CAD) programs that are able to take advantage of the relevant functionality, especially constraint-based modeling. Most robot manufacturers provide accurate 3D CAD models of their robots in a format that is compatible with most CAD systems. The individual links of the robot need to be isolated and treated as separate objects in the CAD environment as well as in the VR environment to which they should be exported in a suitable format, typically Virtual Reality Modeling Language. The links are connected to each other in the VR environment with kinematic joints that are defined exactly as in the real model, thereby forming a kinematic chain. However, even if geometrically accurate models of all equipment elements are available, simulation concerns purely kinematics and does not encompass dynamics and control models, which would have allowed a behavior closer to the real robots. Thus, the robot path derived through a simulator may need to be corrected on the real robot according to calibration and other procedures that may be time consuming, too (Angelidis and Vosniakos, 2014).

Robot simulators based on virtual and augmented reality (VR-AR) were initially simplistic, but more recent developments in VR/AR are increasingly making an impact. Purpose-built VR environments in lieu of previous-generation CAD-based simulators are now commonly available. Multimodal interfaces integrating HMD, haptic devices and force or acceleration sensors have been used in complicated industrial scenarios (Malik et al., 2020; Haton and Mogan, 2008; Mogan et al., 2008). The implementation of such technologies offers the ability to test several algorithms as well as to investigate various signals and means of communication in conjunction with different collaboration schemes, which would be challenging and expensive to conduct in real-life conditions. In a virtual environment various strategies and architectures for safe human–robot collaboration are implemented (Shu et al., 2019; Or et al., 2009); collaborative task processes are optimized (Dombrowski et al., 2017; Tahriri et al., 2015; Wang et al., 2019b); skill acquisition algorithms (Chen et al., 2007) and feedback modalities (Sagardia and Hulin, 2017) are investigated; human–robot interaction techniques are optimized (Gammieri et al., 2017; Matsas et al., 2017). Moreover, path planning is supported in VR/AR by presenting collision-free volumes in VR/AR (Chong et al., 2009), using just a few points to fit trajectory curves through machine learning (Fang et al., 2012), by indicating alternative paths (Hein and Worn, 2009), etc. Also, special interfaces have been developed such as facilitating task recognition through virtual and/or tactile fixtures (Aleotti et al., 2004), using real workpiece data and process limits to effectively define robot operations at the task level through the use of AR (Reinhart et al., 2008), and specifying end-effector orientation by displaying dynamic robot constraints (Fang et al., 2012).

Finally, only a few works focus on methods for robot intent communication with the human (Oyekan et al., 2019; Bolano et al., 2018) as well as the effects of robot characteristics on humans such as its movement speed and path (Koppenborg et al., 2017) and the robot appearance (Weistroffer et al., 2013).

As aforementioned the VR simulation is not only useful for optimizing the HRC design but also for conducting training sessions that can prepare the human workers for real duties or for pre-training when the collaborative task changes. Therefore, numerous researchers have focused their attention on virtual training (Wang and Yang et al., 2011; Roldán et al., 2019; Ji et al., 2018; Jia et al., 2009; Matsas and Vosniakos, 2017; Liu et al., 2013; Al-Ahmari et al., 2016; Nathanael et al., 2016b; Grajewski et al., 2015; Nikolaidis et al., 2015a; Khademian and Hashtrudi-Zaad, 2007; Crespo et al., 2015).

The notable advantage of VR training environments is that they allow the user to interact with a virtual copy of the robot, while they enhance cognition by providing extra information via visual or other perceptual aids or by presenting parts of the real world (as in mixed reality). As aforementioned, these approaches present evident cost and safety benefits in comparison with experimenting in real settings, and most importantly they enhance the information content.

Designing of a Human–Robot Collaborative Task through VR Simulation

The following case study demonstrates the development and testing of an experimental setup for examining alternative HRC schemes (Psarakis et al., 2022). As aforementioned, the combination of human acumen and flexibility with the robot power and precision provides several advantages. In order to investigate the proper and fluent collaboration among these two heterogeneous agents, a simulated industrial task that is part of an assembly line was designed in a VR environment.

In the following sections, the development process and testing of the experimental setup are discussed. First, the task design process is explained (i.e., initial design, iterative improvement testing, and validation of the final design); then the implementation of the virtual environment is presented, followed by the experimental procedure, results obtained, and discussion.

Task Design

Challenges

Designing an appropriate human–robot collaborative task for evaluating HRC presents several challenges. First of all, the designed task should correspond to existing applications while being engaging for the participants, keeping them focused. Additionally, hazardous events such as a collision between the human arm and the robot should be at least somewhat distressing, while the sense of collaboration should be evident enough. Last but not least, the designed task should not take too long to complete to eliminate any fatigue or sickness which may be induced by the prolonged use of HMD (Serge and Moss, 2015). On the contrary, the task duration should be adequate for allowing sufficient data collection. In this particular case study, a human–robot collaborative task was purposefully designed to emulate a generic industrial assembly line/quality control task, where a human has to collaborate with the robot by selecting specific items, leaving the rest to the robot. Specifically the human participant had to gather eighteen balls that were positioned on a table in front of him/her and afterward place them on a conveyor belt that led to a storage basket. The human was in a sitting position with the robot on his/her opposite side. Human participants were instructed to pick a ball once it was highlighted (in Green) while the robot was picking other balls (see Figure 1.1). Assignment of balls to each participant, not visible to them, was actually pre-programmed to make the collaborative task more challenging by increasing the likelihood of collisions between the human and the robot (see Figure 1.2). Two sets of ball trays were processed in each run to allow for adequate task time as well as for a task continuity feeling. Specific parameters like balls' distance in the table, robot speed, picking sequence, threshold for decreasing robot speed, time delays, collision sound, and so on were selected after repeated tests.

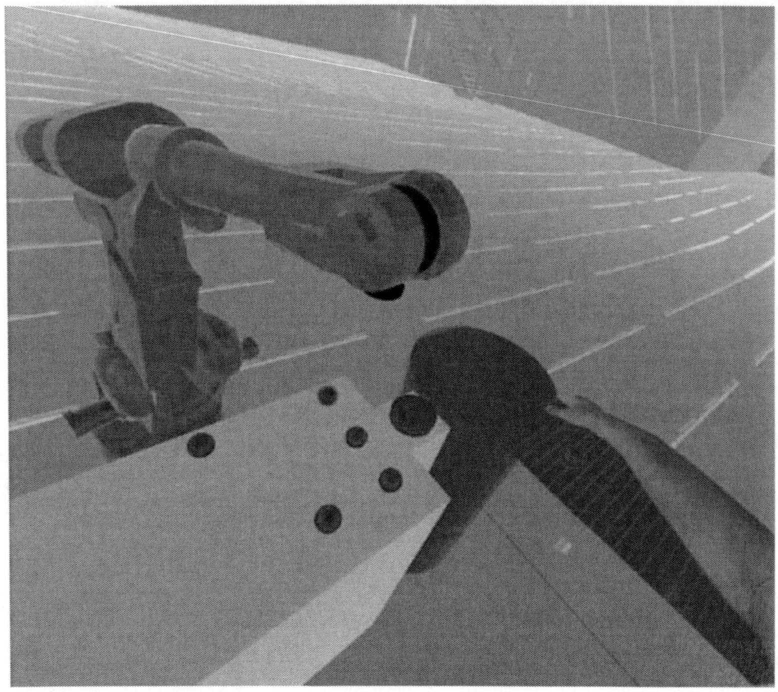

Figure 1.1 The human participant is informed of the next ball to pick by the change of its color to green

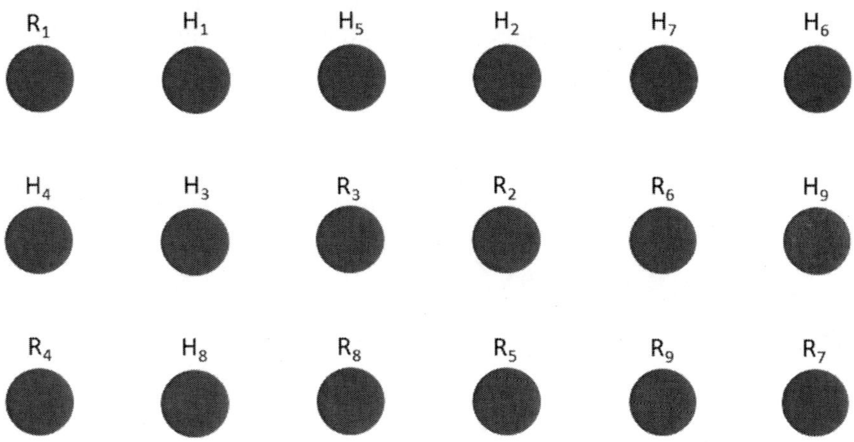

Figure 1.2 The balls' layout along with their picking sequence by the two collaborators (H represents the human while R the robot)

Task Conditions

Three alternative collaborative schemes were designed depending on the robot responsiveness mode:

1. "Base" mode where human participant and robot each collected their assigned balls without provision of any preventive or anticipatory aid.
2. "Prevention" mode where the robot was equipped with a "slow down" function that was set off when the distance between the virtual hand and itself was less than twenty centimeters.

The "Slow down" function reduces the robot speed by 50% for preventing possible collisions by increasing the time gap of human reaction. When the robot detects proximity greater than twenty centimeters, the "slow down" function stops and the robot speed is restored.

3. "Anticipation" mode where the use of orange color signifies the robot's next ball (see Figure 1.3). This indication was explicitly designed to help the human participant anticipate the robot's next movement and avoid collision.

Iterative Improvement Tests

Starting with the original configuration several pilot runs were performed, iteratively tweaking and redesigning parts of the experimental task. Some of the issues identified and their solutions are described below.

Even though the instructed participants' primary goal was task accomplishment (i.e., time and accuracy) under the umbrella of safety (i.e., collision avoidance), it was noticed by the experimenters that the impact of collision alarm was so great that certain pilot participants tended to adopt robot avoidance as their primary concern. Hence, the collision auditory and visual alarm signals were attenuated to better correspond to the criticality level of such events.

Moreover, a sense of a rather competitive task was noticed among the participants. Typical commentaries included "I pick up my objects and the robot picks up its own" and "I need to finish first before the robot does." For that reason, a task redesign included the control of the pace by the human participant; that is, the robot was unable to surpass the participant. Furthermore, in case of a collision the robot ceased its operation until the human pressed a button.

Experimenters also noticed that a number of participants tended to put their hands below the table, avoiding the robot in this way. To this end at the beginning of each run the height of

Figure 1.3 The human participant is informed of the next ball the robot is about to pick by the change of its color to orange

the virtual table was adjusted to fit exactly with the real one, depending on the anthropometrics of each participant, thus making the simulation more realistic and preventing any "illegal" movements by participants.

Likewise, in their first run—regardless of condition—participants spent some time getting used to their virtual hand manipulation. Moreover, when their first collision with the robot occurred, there was a delay in the procedure needed for resuming the task (i.e., pressing the button). So a practice tutorial was added to help participants familiarize with the virtual environment and their virtual hand movement in order to reduce learning time in their first run and to counterbalance the order effect.

Finally yet importantly, when positioning a ball to the conveyor belt, some participants felt like the ball teleported back to the table (as in that moment the next ball they had to pick was indicated with the corresponding green color). Redesign involved adding a time delay of 200 msec before highlighting the next ball to be picked.

Final Design

Several pilot tests were conducted to evaluate the adequacy of the experimental testbed for studying the aforementioned HRC schemes. In Figure 1.4 a trial with the final design is displayed.

Apparatus

In this section the hardware and software components of the system used for the implementation of the virtual collaborative task are described. A wearable tracking system and a HMD were used for manipulating the virtual avatar and for stereoscopic visual display respectively. In Figure 1.5 an illustration of the system is displayed. An external camera was used for recording the experiment trials in order to identify any procedure mistakes and for post analysis.

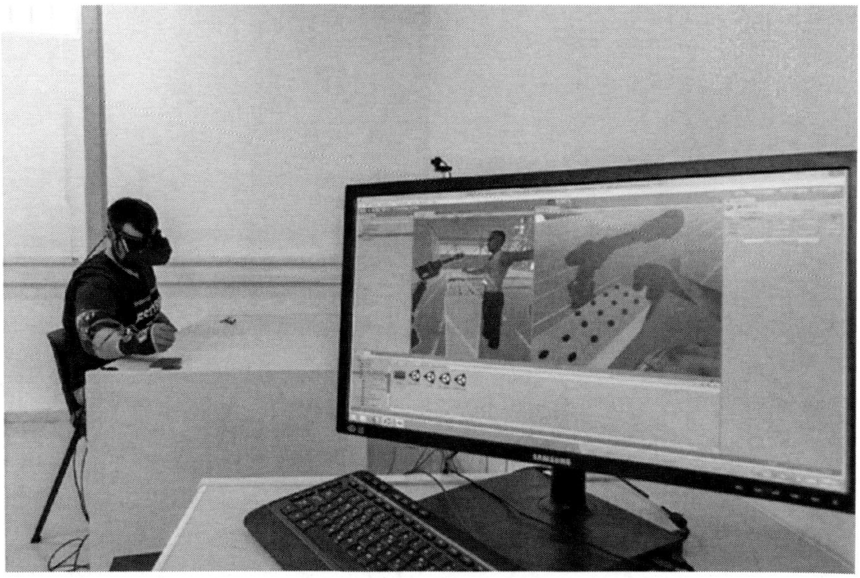

Figure 1.4 The final experimental setup where a subject carries out the designed task while the experimenter observes in real time the virtual simulation

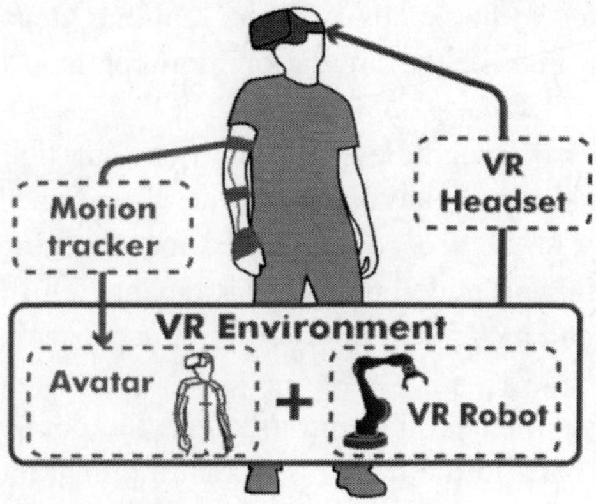

Figure 1.5 Overview of the technical system

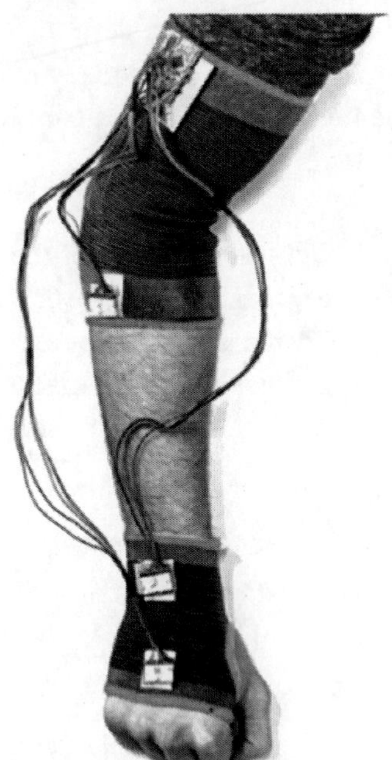

Figure 1.6 Arm mount of the current motion-tracking system

Inertial Measurement Unit (IMU) Wearable Tracking System

A motion-tracking system capable of tracking the movement of the human arm was used (Mourelatos et al., 2019). The wearable system includes a fingerless glove and an elbow patch that are attached to an Arduino Nano and three MPU 9250 IMUs. Its main advantage over the existing arm tracking systems is its portability as it can operate independently from any type of position-tracking technology or display. In Figure 1.6 the wearable motion-tracking system is displayed.

HMD

The Oculus Rift DK2 Virtual Reality HMD was used as an output device for visual display. The Oculus Rift offers a 100° field of view, frame rate of 60 fps, and resolution of 1920 × 1080, resulting in participants' immersion increase. The movement of participants' head is tracked in real time by a sensor camera. For minimizing motion sickness the leaning tracking sensor was deactivated.

Implementation of the Virtual Environment

The designed virtual environment was developed in the cross-platform game engine Unity 3D™. Unity 3D is a popular game development platform, which is used to create both 3D and 2D games and interactive simulations. All the necessary resources for the creation of the simulation are imported to Unity as assets (such as graphic object models, materials, textures, and sounds), along with the written scripts that are essential for the proper and interactive simulation (such as triggering mechanisms, time events, avatar movement, events, collision detection, and kinematic behaviors). All the aforementioned components are used for the creation of a final scene where a human subject can navigate. The structure of the Unity user interface is shown in Figure 1.7.

Virtual Robot

The robot is an ABB IRB 6600 industrial robot modeled in Blender™ and rigged. Rigging is a technique for creating the bone structure of a 3D model by interconnecting them forming a hierarchy. The inverse kinematics of the current model was used, so the robot's movement inside the VE is natural and smooth.

The movement of the robot was created through animation and consisted of (i) the idle state (i.e. its initial position), (ii) the picking and placement process, and (iii) the return to the initial position (see, e.g., Figure 1.8). For every pick, the animation procedure lasted about 180 frames

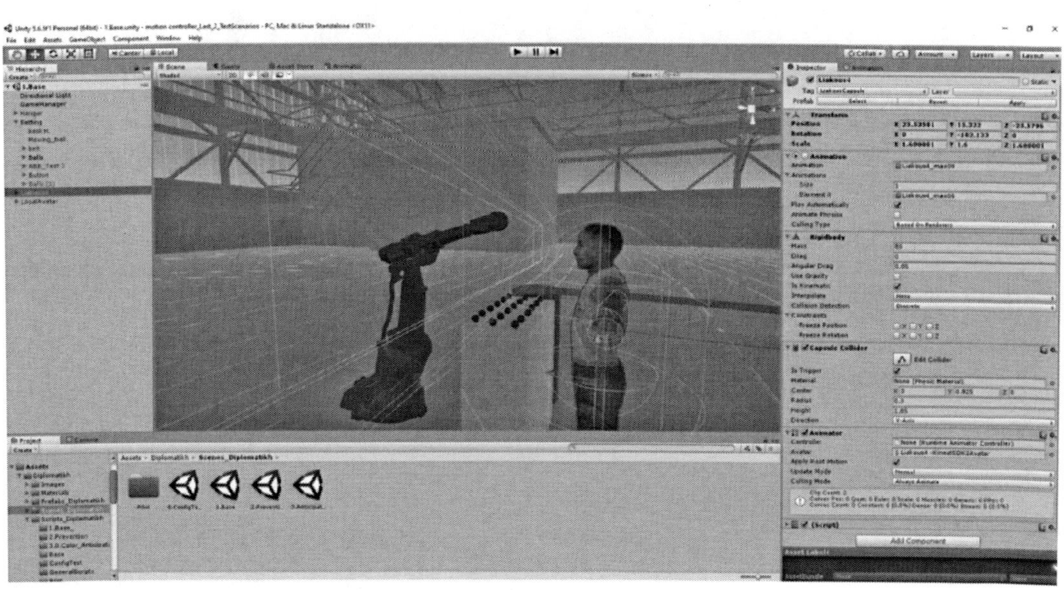

Figure 1.7 Overview of the environment of a Unity project

Figure 1.8 Animation of the robot picking and placing a ball

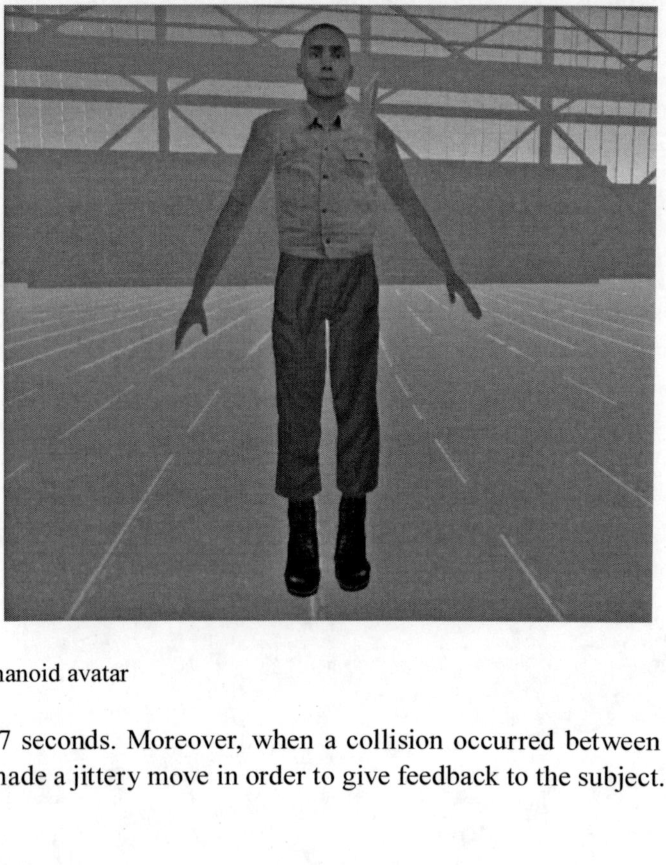

Figure 1.9 The humanoid avatar

or approximately 7 seconds. Moreover, when a collision occurred between the collaborating agents, the robot made a jittery move in order to give feedback to the subject.

Human Avatar

The human avatar (Figure 1.9) is a humanoid model made by Matsas et al. (2012). The model was created with the online tool Evolver^{TM}, and afterward it was imported to 3ds Max^{TM} where a biped kinematic skeleton and basic postures, that is, "T" and "Y" posture, were adopted. In the right hand of the avatar, a script regarding the use of the wearable tracking system was added.

Work Cell Setup

The remaining parts of the work cell include the table where the balls are placed, the button that is used for resuming the process, and the conveyor belt where the balls are disposed

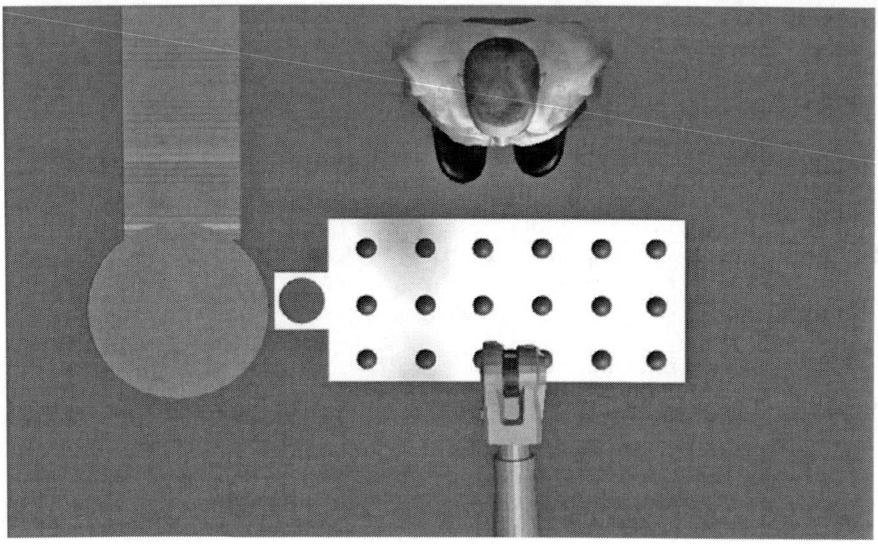

Figure 1.10 Setup of the virtual collaborating task

of by the collaborating agents. The human and the robot are located opposite to each other and share the same workspace, i.e., the table with the balls. The work cell is stationed inside an industrial hangar. The game objects' scale and position are designed so that both agents are able to reach all the balls, while the robot's end effector does not get too close to the human's face to avoid a feeling of fear. An overview of the virtual setup is shown in Figure 1.10.

Experimental Design

In order to reduce individual differences bias a within-subject design was preferred for studying the effects of robot visual anticipatory cues and of robot preventive reactions on the designed task. The independent variable was the robot mode and includes three conditions Control (C), Prevention (P), and Anticipation (A). Furthermore, to account for learning or order effects, the conditions order was fully compensated. Every participant carried out the simulated task in all conditions.

Participants

Thirty-three male and fifteen female university students participated voluntarily in the trials. Their mean age was 24.5 years old with a standard deviation of 2.5 years.

Metrics

Three dependent variables were used for measuring the task performance, human safety, and collaboration fluency. Specifically, the number of collisions, the task completion time, and the human idle time (the time the participant was inactive) were recorded and analyzed.

Experimental Procedure

Prior to the beginning of the trials, verbal informed consent was obtained from the participants and the experimenter gave them instructions regarding the task procedure and goal. Afterward participants were placed in a sitting position in front of a table. At the start of each condition additional information was provided to them and a calibration of the wearable tracking system was executed. After the completion of the experiment, participants filled out a questionnaire mainly focusing on their general experience inside the virtual environment.

Results

The IBM SPSS Statistics tool was used for analyzing the collected data. Summary data charts are presented in Figure 1.11. A more comprehensive presentation of the results can be found in Psarakis et al. (2022).

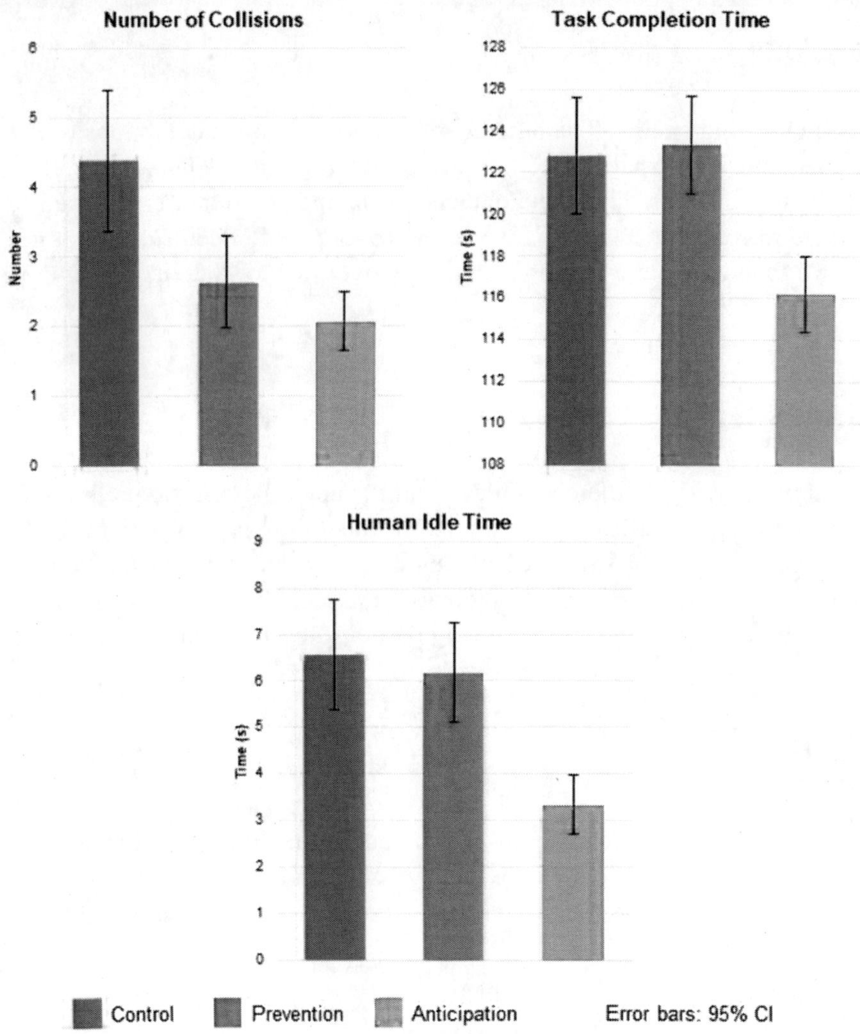

Figure 1.11 The mean values of the number of collisions (top left), the task completion time (top right), and the human idle time (bottom) for all three conditions (error bars of 95% CI).

Number of Collisions

A significant main effect of robot mode on the number of collisions was found (χ^2 (2) = 17.354, $p < .001$) (Table 1.1). Post hoc analysis showed that participants made significantly fewer collisions in the Anticipation condition compared to the Control condition ($Z = -4.008, p < .001$) as well as in the Prevention condition compared to the Control condition ($Z = -2.940, p = .003$). No statistically significant difference was found between the Anticipation and the Prevention condition ($Z = -1.563, p = .118$).

Task Completion Time

There was a statistically significant difference in the task completion time (F (2, 88) = 12.205, $p < .001$, $\eta_p^2 = .217$) (Table 1.2). Post hoc analysis showed that participants finished the task significantly earlier in the Anticipation condition compared to the Prevention condition ($p < .001$) and to the Control condition (p = .001). Nevertheless, no significant difference was found between the Prevention and the Control condition ($p > .05$).

Human Idle Time

There was a statistically significant difference in the total human idle time (f (2, 88) = 26.064, $p < .001$, $\eta_p^2 = .372$) (Table 1.3). Post hoc analysis showed that participants were idle for significantly less time in the Anticipation condition compared to the Prevention condition ($p < .001$) and to the Control condition ($p < .001$). Nevertheless, no significant difference was found between the Prevention and the Control condition ($p > .05$).

Table 1.1 Number of collisions

Control		Prevention		Anticipation				
M	SD	*M*	SD	*M*	SD	χ^2	*df*	*p*
4.38	3.40	2.64	2.20	2.07	1.42	17.354	2	< .001

Table 1.2 Task completion time

Control		Prevention		Anticipation				
M	SD	*M*	SD	*M*	SD	*F*	η_p^2	*p*
122.80	9.23	123.33	7.69	116.20	6.06	12.205	.217	< .001

Table 1.3 Human idle time

Control		Prevention		Anticipation				
M	SD	*M*	SD	*M*	SD	*F*	η_p^2	*p*
6.57	3.97	6.18	3.60	3.34	2.11	26.064	.372	< .001

Subjective Evaluation

The subjective rating data from the questionnaire supported the effectiveness of the VR simulation, regarding both the task design and the user experience. In particular, 90% of the subjects felt the immersion of the VE, while 75% experienced the feeling of actually grabbing a physical ball. Moreover, 94% did not experience hand movement restrictions whereas 83% reported natural hand movement. Furthermore, 92% reported awareness when a collision occurred, while nearly no one, that is, only 2%, felt any kind of discomfort and only 21% of the subjects did not feel that the robot was collaborating with them.

Last but not least, approximately 70% of the subjects answered that they would feel safer sharing a real workspace with a robot in the Anticipation condition and the remaining 30% preferred the Prevention condition. The Control condition did not receive any selection.

Discussion

The case study presented earlier made use of a VR simulator to examine the pros and cons of two alternative HRC schemes for a particular industrial sorting task. The results obtained clearly support that, at least in principle, providing timely cues to the human as to the robot's imminent move was the most promising option in terms of both efficiency and safety.

Specifically, in terms of collisions, both Anticipation and Prevention conditions clearly outperformed Control condition with significantly fewer collisions occurring. Differences between the latter two were not found to be significant. In terms of both task time and human idle time, Anticipation condition clearly outperformed both Prevention and Control conditions with the difference between the latter two not being significant.

The following design recommendations can be derived from the above findings:

An HR collaboration scheme that incorporates a reduction of robot speed when sensing human body parts close to the robot arm (Prevention condition) results in fewer collisions, thus improving the safety of the human operator without negatively affecting task efficiency.

An HR collaboration scheme that provides anticipatory cues of the robot's next move to the human operator (anticipation condition) results (i) in fewer collisions thus improving operator safety, (ii) in lower task time thus also improving HR team productivity, and (iii) in lower human operator idle time thus increasing human efficiency.

Subjective evaluation was largely in accordance with the above findings with most participants preferring the Anticipation condition. One however should be cautioned on applying these recommendations directly to a real HRC workstation without prior pilot testing. In a real situation workers will have much more time to be acquainted with the robot's behavior and are also expected to be more cautious regarding possible collisions. This is due to the fact that in a virtual environment the perceived feeling of safety may be misinterpreted by the users and therefore encourage them to take bigger risks comparing to working beside a real robot in a real work cell.

Notwithstanding the aforementioned cautions, evidently the same simulator, with minor alterations can serve as a testbed for various other options (e.g., a combined Prevention/Anticipation scheme or a human intent recognition algorithm). In addition, various other collaboration parameters can be readily examined such as work cell layout, robot speed, or alternative robot arm trajectories to optimize the final design.

Conclusion

The present chapter provided a review of contemporary work and a specific case study on the use of VR simulators for the design of human–robot collaborative workstations. Several challenges remain in employing VR simulators for HRC workstations, such as a lack of realism due to limited or unnatural tactile and kinesthetic sensing, possible dizziness, or other unpleasant symptoms. However, the use of VR simulators, if properly developed, through iterative design cycles to tune the experiential resemblance of the virtual task to the real one, is a promising, swift, and cost-effective method for testing design options for HR collaborative workstations.

References

Al-Ahmari, A. M., Abidi, M. H., Ahmad, A., & Darmoul, S. (2016). Development of a virtual manufacturing assembly simulation system. *Advances in Mechanical Engineering*, 8(3), 1687814016639824.

Aleotti, J., Caselli, S., & Reggiani, M. (2004). Leveraging on a VE for robot programming by demonstration. *Robotics and Autonomous Systems*, 47(2), 153–161.

Angelidis, A., & Vosniakos, G.-C. (2014). Prediction and compensation of relative position error along industrial robot end-effector paths. *International Journal of Precision Engineering and Manufacturing*, 15(1), 66–73.

Aurich, J. C., Ostermayer, D., & Wagenknecht, C. H. (2009). Improvement of manufacturing processes with virtual reality-based CIP workshops. *International Journal of Production Research*, 47(19), 5297–5309.

Baraka, K., Paiva, A., & Veloso, M. (2016). Expressive lights for revealing mobile service robot state. In *Robot 2015: Second Iberian Robotics Conference* (pp. 107–119). Springer, Cham.

Bardy, B. G., Lagarde, J., & Mottet, D. (2012). Dynamics of skill acquisition in multimodal technological environments. Chapter 3 in *Skill Training in Multimodal Virtual Environments*, edited by Bergamasco, M., Bardy, B. G., & Gopher, D. (pp. 31–45). London: CRC Press.

Bergamasco, M., Bardy, B. G., & Gopher, D., eds. (2012). *Skill Training in Multimodal Virtual Environments*. London: CRC Press.

Bolano, G., Roennau, A., & Dillmann, R. (2018). Transparent robot behavior by adding intuitive visual and acoustic feedback to motion replanning. In *2018 27th IEEE International Symposium on Robot and Human Interactive Communication (RO-MAN)* (pp. 1075–1080). IEEE, August.

Brooks, B. M. (1999). The specificity of memory enhancement during interaction with a virtual environment. *Memory*, 7(1), 65–78.

Chen, Y., Han, X., Okada, M., Chen, Y., & Naghdy, F. (2007). Intelligent robotic peg-in-hole insertion learning based on haptic virtual environment. In *2007 10th IEEE International Conference on Computer-Aided Design and Computer Graphics* (pp. 355–360). IEEE, October.

Chong, J. W. S., Ong, S. K., Nee, A. Y. C., & Youcef-Youmi, K. (2009). Robot programming using augmented reality: An interactive method for planning collision-free paths. *Robotics and Computer-Integrated Manufacturing*, 25(3), 689–701.

Crespo, R., García, R., & Quiroz, S. (2015). Virtual reality application for simulation and off-line programming of the mitsubishi movemaster RV-M1 robot integrated with the oculus rift to improve students training. *Procedia Computer Science*, 75, 107–112.

Dankelman, J., Wentink, M., & Stassen, H. G. (2003). Human reliability and training in minimally invasive surgery. *Minimally Invasive Therapy and Allied Technologies*, 12, 1229–1235.

de Freitas, F. V., Gomes, M. V. M., & Winkler, I. (2022). Benefits and challenges of virtual-reality-based industrial usability testing and design reviews: A patents landscape and literature review. *Applied Sciences*, 12(3), 1755.

de Gea Fernández, J., Mronga, D., Günther, M., Knobloch, T., Wirkus, M., Schröer, M., ... Kirchner, F. (2017). Multimodal sensor-based whole-body control for human–robot collaboration in industrial settings. *Robotics and Autonomous Systems*, 94, 102–119.

de Giorgio, A., Romero, M., Onori, M., & Wang, L. (2017). Human-machine collaboration in virtual reality for adaptive production engineering. *Procedia Manufacturing*, 11, 1279–1287. https://doi.org/10.1016/j.promfg.2017.07.255.

Dimitrokalli, A., Vosniakos, G. C., Nathanael, D., & Matsas, E. (2020). On the assessment of human-robot collaboration in mechanical product assembly by use of Virtual Reality. *Procedia Manufacturing*, 51, 627–634.

Dombrowski, U., Stefanak, T., & Perret, J. (2017). Interactive simulation of human-robot collaboration using a force feedback device. *Procedia Manufacturing*, 11, 124–131.

Faccio, M., Minto, R., Rosati, G., & Bottin, M. (2020). The influence of the product characteristics on human-robot collaboration: A model for the performance of collaborative robotic assembly. *The International Journal of Advanced Manufacturing Technology*, 106(5), 2317–2331.

Fang, H. C., Ong, S. K., & Nee, A. Y. C. (2012). Interactive robot trajectory planning and simulation using augmented reality. *Robotics and Computer Integrated Manufacturing*, 28(2), 227–238.

Gallagher, A. G., McClure, N., McGuigan, J., Ritchie, K., & Shechy, N. P. (1998). An ergonomic analysis of the 'fulcrum effect' in endoscopic skill acquisition. *Endoscopy*, 30, 617–620.

Gallagher, A. G., McClure, N., McGuigan, J., Crothers, I., & Browning, J. (1999). Virtual reality training in laparoscopic surgery: A preliminary assessment of minimally invasive surgical trainer virtual reality (MIST VR). *Endoscopy*, 4, 310–313.

Gammieri, L., Schumann, M., Pelliccia, L., Di Gironimo, G., & Klimant, P. (2017). Coupling of a redundant manipulator with a virtual reality environment to enhance human-robot cooperation. *Procedia CIRP*, 62, 618–623.

Grajewski, D., Górski, F., Hamrol, A., & Zawadzki, P. (2015). Immersive and haptic educational simulations of assembly workplace conditions. *Procedia Computer Science*, 75, 359–368.

Gupta, S. K., Anand, D. K., Brough, J., Schwartz, M., & Kavetsky, R. (2008). Training in virtual environments. In *A Safe, Cost-Effective, and Engaging Approach to Training*. CALCE EPSC Press, University of Maryland, College Park, MD.

Haton, B., & Mogan, G. (2008). Enhanced ergonomics and virtual reality applied to industrial robot programming. In *Proceedings 4th International Conference on Robotics* (pp. 595–604). Brasov, Romania, 13–14 November.

Heilig, M. (1962). *Sensorama Simulator. U.S. Patent No – 3, 870*. Virginia: United States Patent and Trade Office.

Hein, B., & Worn, H. (2009). Intuitive and model-based on-line programming of industrial robots: New input devices. In *Proceedings of the IEEE/RSJ International Conference on Intelligent Robots and Systems (IROS 2009)* (pp. 3064–3069), 10–15 October.

Hoffman, G. (2019). Evaluating fluency in human–robot collaboration. *IEEE Transactions on Human-Machine Systems*, 49(3), 209–218.

Huang, C. M., & Mutlu, B. (2016). Anticipatory robot control for efficient human-robot collaboration. In *2016 11th ACM/IEEE International Conference on Human-Robot Interaction (HRI)* (pp. 83–90). IEEE, March.

Huckaby, J., & Christensen, H. I. (2012). *A Taxonomic Framework for Task Modeling and Knowledge Transfer in Manufacturing Robotics Association for the Advancement of Artificial Intelligence Technical Report WS-12-06* (pp. 95–101).

International Federation of Robotics. (2021a). *World Robotics 2021 – Industrial Robots*, IFR Statistical Department, VDMA Services GmbH, Frankfurt am Main, Germany, 2021.

International Federation of Robotics. (2021b). *World Robotics 2021 – Service Robots*, IFR Statistical Department, VDMA Services GmbH, Frankfurt am Main, Germany, 2021.

Ji, W., Yin, S., & Wang, L. (2018). A virtual training based programming-free automatic assembly approach for future industry. *IEEE Access*, 6, 43865–43873.

Jia, D., Bhatti, A., & Nahavandi, S. (2009). Design and evaluation of a haptically enable virtual environment for object assembly training. In *2009 IEEE International Workshop on Haptic Audio visual Environments and Games* (pp. 75–80). IEEE, November.

Khademian, B., & Hashtrudi-Zaad, K. (2007). Performance issues in collaborative haptic training. In *Proceedings 2007 IEEE International Conference on Robotics and Automation* (pp. 3257–3262). IEEE, April.

Kim, W., Lorenzini, M., Balatti, P., Nguyen, P. D., Pattacini, U., Tikhanoff, V., ... Ajoudani, A. (2019). Adaptable workstations for human-robot collaboration: A reconfigurable framework for improving worker ergonomics and productivity. *IEEE Robotics & Automation Magazine*, 26(3), 14–26.

Koppenborg, M., Nickel, P., Naber, B., Lungfiel, A., & Huelke, M. (2017). Effects of movement speed and predictability in human–robot collaboration. *Human Factors and Ergonomics in Manufacturing & Service Industries*, 27(4), 197–209.

Lasota, P. A., Fong, T., & Shah, J. A. (2017). A survey of methods for safe human-robot interaction. *Foundations and Trends® in Robotics*, 5(4), 261–349.

Lin, F., Lan, Y., Duffy, V., & Su, A. (2002). Developing virtual environments for industrial training. *Information Sciences-Informatics and Computer Science: International Journal*, 140(1), 153–170.

Liu, H., & Wang, L. (2018). Gesture recognition for human-robot collaboration: A review. *International Journal of Industrial Ergonomics*, 68, 355–367.

Liu, P., Glas, D. F., Kanda, T., & Ishiguro, H. (2018). Learning proactive behavior for interactive social robots. *Autonomous Robots*, 42(5), 1067–1085.

Liu, X. H., Song, G. M., Cui, X. L., Xu, B. H., Wang, K., & Wang, Z. B. (2013). Development of a virtual maintenance system for complex mechanical product. *Advances in Mechanical Engineering*, 5, 730925.

Malik, A. A., Masood, T., & Bilberg, A. (2020). Virtual reality in manufacturing: Immersive and collaborative artificial-reality in design of human-robot workspace. *International Journal of Computer Integrated Manufacturing*, 33(1), 22–37.

Marmaras, N., Vassilakopoulou, P., & Salvendy, G. (1997). Developing a cognitive aid for CNC lathe programming through problem-driven approach. *International Journal of Cognitive Ergonomics*, 1(3), 267–289.

Matsas, E., & Vosniakos, G. C. (2017). Design of a virtual reality training system for human–robot collaboration in manufacturing tasks. *International Journal on Interactive Design and Manufacturing (IJIDeM)*, 11(2), 139–153.

Matsas, E., Batras, D., & Vosniakos, G. C. (2012). Beware of the robot: A highly interactive and immersive virtual reality training application in robotic manufacturing systems. In *IFIP International Conference on Advances in Production Management Systems* (pp. 606–613). Springer, Berlin, Heidelberg, September.

Matsas, E., Vosniakos, G. C., & Batras, D. (2017). Effectiveness and acceptability of a virtual environment for assessing human–robot collaboration in manufacturing. *The International Journal of Advanced Manufacturing Technology*, 92(9), 3903–3917.

Matsas, E., Vosniakos, G. C., & Batras, D. (2018). Prototyping proactive and adaptive techniques for human-robot collaboration in manufacturing using virtual reality. *Robotics and Computer-Integrated Manufacturing*, 50, 168–180.

Maurtua, I., Ibarguren, A., Kildal, J., Susperregi, L., & Sierra, B. (2017). Human–robot collaboration in industrial applications: Safety, interaction and trust. *International Journal of Advanced Robotic Systems*, 14(4), 1729881417716010.

Michalos, G., Makris, S., Tsarouchi, P., Guasch, T., Kontovrakis, D., & Chryssolouris, G. (2015). Design considerations for safe human-robot collaborative workplaces. *Procedia CIRP*, 37, 248–253.

Mogan, G., Talaba, D., Girbacia, F., Butnaru, T., Sisca, S., & Aron, C. (2008). A generic multimodal interface for design and manufacturing applications. In *Proceedings of the 2nd International Workshop Virtual Manufacturing (VirMan08) - Part of the 5th Intuition International Conference: Virtual Reality in Industry and Society: From Research to Application* (p. 10). Torino, Italy, 6–8 October.

Mourelatos, A., Nathanael, D., Gkikas, K., & Psarakis, L. (2018). Development and evaluation of a wearable motion tracking system for sensorimotor tasks in VR environments. In *Congress of the International Ergonomics Association* (pp. 181–188). Springer, Cham, August.

Müller-Abdelrazeq, S. L., Schönefeld, K., Haberstroh, M., & Hees, F. (2019). Interacting with collaborative robots—a study on attitudes and acceptance in industrial contexts. In *Social Robots: Technological, Societal and Ethical Aspects of Human-Robot Interaction* (pp. 101–117). Springer, Cham.

Nathanael, D., Mosialos, S., & Vosniakos, G. C. (2016a). Development and evaluation of a virtual training environment for on-line robot programming. *International Journal of Industrial Ergonomics*, 53, 274–283.

Nathanael, D., Mosialos, S., Vosniakos, G. C., & Tsagkas, V. (2016b). Development and evaluation of a virtual reality training system based on cognitive task analysis: The case of CNC tool length offsetting. *Human Factors and Ergonomics in Manufacturing & Service Industries*, 26(1), 52–67.

Nikolaidis, S., Lasota, P., Ramakrishnan, R., & Shah, J. (2015a). Improved human–robot team performance through cross-training, an approach inspired by human team training practices. *The International Journal of Robotics Research*, 34(14), 1711–1730.

Nikolaidis, S., Ramakrishnan, R., Gu, K., & Shah, J. (2015b). Efficient model learning from joint-action demonstrations for human-robot collaborative tasks. In *2015 10th ACM/IEEE International Conference on Human-Robot Interaction (HRI)* (pp. 189–196). IEEE.

Oman, C. M. (1982). A heuristic mathematical model for the dynamics of sensory conflict and motion sickness. *Acta Oto-Laryngologica*, 94(Suppl 392), 4–44.

Ong, S. K., Yuan, M. L., & Nee, A. Y. C. (2008). Augmented reality applications in manufacturing: A survey. *International Journal of Production Research*, 46(10), 2707–2742.

Or, C. K., Duffy, V. G., & Cheung, C. C. (2009). Perception of safe robot idle time in virtual reality and real industrial environments. *International Journal of Industrial Ergonomics*, 39(5), 807–812.

Oyekan, J. O., Hutabarat, W., Tiwari, A., Grech, R., Aung, M. H., Mariani, M. P., ... Dupuis, C. (2019). The effectiveness of virtual environments in developing collaborative strategies between industrial robots and humans. *Robotics and Computer-Integrated Manufacturing*, 55, 41–54.

Psarakis, L., Nathanael, D., & Marmaras, N. (2022). Fostering short-term human anticipatory behavior in human-robot collaboration. *International Journal of Industrial Ergonomics*, 87, 103241.

Rasmussen, J. (1983). Skills, rules, and knowledge; signals, signs, and symbols, and other distinctions in human performance models. *IEEE Transactions on Systems, Man, and Cybernetics*, 3, 257–266.

Reason, J. T., & Brand, J. J. (1975). *Motion Sickness*. Academic Press, Oxford, England.

Regian, J. W. (1997). Virtual Reality for Training: Evaluating Transfer. Chapter 3 in *Virtual Reality, Training's Future?*. Defense Research Series, vol 6. edited by Seidel, R.J., Chatelier, P.R. Springer, Boston, MA. https://doi.org/10.1007/978-1-4899-0038-8_4.

Reinhart, G., Munzert, U., & Vogl, W. (2008). A programming system for robot-based remote-laser-welding with conventional optics. *CIRP Annals - Manufacturing Technology*, 57, 37–40.

Riccio, G. E., & Stoffregen, T. A. (1991). An ecological theory of motion sickness and postural instability. *Ecological Psychology*, 3(3), 195–240.

Robla-Gómez, S., Becerra, V. M., Llata, J. R., Gonzalez-Sarabia, E., Torre-Ferrero, C., & Perez-Oria, J. (2017). Working together: A review on safe human-robot collaboration in industrial environments. *IEEE Access*, 5, 26754–26773.

Roldán, J. J., Crespo, E., Martín-Barrio, A., Peña-Tapia, E., & Barrientos, A. (2019). A training system for Industry 4.0 operators in complex assemblies based on virtual reality and process mining. *Robotics and Computer-Integrated Manufacturing*, 59, 305–316.

Rose, F. D., Attree, E. A., Brooks, B. M., Parslow, D. M., & Penn, P. R. (2000). Training in virtual environments: Transfer to real world tasks and equivalence to real task training. *Ergonomics*, 43(4), 494–511.

Rückert, P., Wohlfromm, L., & Tracht, K. (2018). Implementation of virtual reality systems for simulation of human-robot collaboration. *Procedia Manufacturing*, 19, 164–170.

Sagardia, M., & Hulin, T. (2017). Multimodal evaluation of the differences between real and virtual assemblies. *IEEE Transactions on Haptics*, 11(1), 107–118.

Sanchez-Vives, M. V., & Slater, M. (2005). From presence to consciousness through virtual reality. *Nature Reviews Neuroscience*, 6(4), 332–339.

Serge, S. R., & Moss, J. D. (2015). Simulator sickness and the oculus rift: A first look. In *Proceedings of the Human Factors and Ergonomics Society Annual Meeting* (Vol. 59, No. 1, pp. 761–765). SAGE Publications, Sage CA, Los Angeles, CA, September.

Shu, B., Sziebig, G., & Pieters, R. (2019). Architecture for safe human-robot collaboration: Multi-modal communication in virtual reality for efficient task execution. In *2019 IEEE 28th International Symposium on Industrial Electronics (ISIE)* (pp. 2297–2302). IEEE, June.

Slater, M., & Wilbur, S. A. (1997). Framework for Immersive Virtual Environments (FIVE): Speculations on the role of presence in virtual environments. *Presence: Teleoperators and Virtual Environments*, 6(6), 603–616.

Sutherland, I. E. (1965). The ultimate display. In *Proceedings of the IFIP Congress* (Vol. 2, pp. 506-508). Spartan Books, Washington, D.C., May.

Sutherland, I. E. (1968, December). A head-mounted three dimensional display. In *Proceedings of the December 9-11, 1968, Fall Joint Computer Conference, Part I* (pp. 757–764). Association for Computing Machinery, New York, NY, USA. https://doi.org/10.1145/1476589.1476686.

Tahriri, F., Mousavi, M., Yap, H. J., Siti Zawiah, M. D., & Taha, Z. (2015). Optimizing the robot arm movement time using virtual reality robotic teaching system. *International Journal of Simulation Modelling*, 14(1), 28–38.

Takayama, L., Ju, W., & Nass, C. (2008). Beyond dirty, dangerous and dull: What everyday people think robots should do. In *2008 3rd ACM/IEEE International Conference on Human-Robot Interaction (HRI)* (pp. 25–32). IEEE, March.

Waller, D., Hunt, E., & Knapp, D. (1998). The transfer of spatial knowledge in virtual environment training. *Presence: Teleoperators and Virtual Environments*, 7(2), 129–143.

Wang, L., & Yang, X. (2011). Assembly operator training and process planning via virtual systems. *International Journal of Sustainable Engineering*, 4(1), 57–67.

Wang, L., Gao, R., Váncza, J., Krüger, J., Wang, X. V., Makris, S., & Chryssolouris, G. (2019a). Symbiotic human-robot collaborative assembly. *CIRP Annals*, 68(2), 701–726.

Wang, L., Liu, S., Liu, H., & Wang, X. V. (2020). Overview of human-robot collaboration in manufacturing. In *Proceedings of 5th International Conference on the Industry 4.0 Model for Advanced Manufacturing* (pp. 15–58). Springer, Cham.

Wang, Q., Cheng, Y., Jiao, W., Johnson, M. T., & Zhang, Y. (2019b). Virtual reality human-robot collaborative welding: A case study of weaving gas tungsten arc welding. *Journal of Manufacturing Processes*, 48, 210–217.

Weiss, A., Igelsböck, J., Wurhofer, D., & Tscheligi, M. (2011). Looking forward to a "robotic society"? *International Journal of Social Robotics*, 3(2), 111–123.

Weistroffer, V., Paljic, A., Callebert, L., & Fuchs, P. (2013). A methodology to assess the acceptability of human-robot collaboration using virtual reality. In *Proceedings of the 19th ACM Symposium on Virtual Reality Software and Technology* (pp. 39–48). October.

Wentink, M., Stassen, L. P. S., Alwayn, I., Hosman, R. J. A. W., & Stassen, H. G. (2003). Rasmussen's model of human behavior in laparoscopy training. *Surgical Endoscopy and Other Interventional Techniques*, 17(8), 1241–1246.

Wood, D. J., & Gray, B. (1991). Toward a comprehensive theory of collaboration. *The Journal of Applied Behavioral Science*, 27, 139–162. https://doi.org/10.1177/0021886391272001.

Chapter 2

Introduction to BIM for Heritage

Demitris Galanakis,[1] Danae Phaedra Pocobelli,[1]
Antonios Konstantaras,[1] Katerina Mania,[2] Emmanuel Maravelakis[1]

[1] Department of Electronic Engineering, Hellenic Mediterranean University, Chania, Greece
[2] Department of Electrical and Computer Engineering, Technical University of Crete, Chania, Greece

Abstract

This chapter defines the Building Information Modelling (BIM) methodology and explores its potential for Cultural Heritage and Conservation. Next, the discussion addresses reasons behind foreseen BIM increase in popularity and applicability across new architectural fields, such as heritage-specific applications. Obstacles and open issues in generalising BIM for Heritage exhibits are analysed next, with emphasis being given to "level of maturity" and "data diversity". Scan-to-BIM is addressed immediately after, and a short description of each entailed step follows – starting from data acquisition until "smart-object" finalisation and metadata handling. Finally, we list literature attempts to apply *Historic Building Information Modelling* (HBIM) technology in several fields, from parametric libraries created *ad hoc* for specific architecture styles, such as Murphy's HBIM and Baik's *Jeddah Historic Building Information Modelling*, to dedicated usage in *Facility Management*. Analysed endeavours include seismic effect mitigation, structural analysis, environmental monitoring/modelling, and integration with *Geographic Information System* (GIS), in order to outline HBIM usage potential.

Keywords: 3D point cloud data, BIM, HBIM, JHBIM, mesh segmentation, 3D modelling, Scan-to-BIM, deep learning, machine learning, Reconstruction Information Modelling, Digital Twin, Hybrid BIM, hyperspectral data

From BIM to Heritage-Dedicated BIM (HBIM)

Introduction

This chapter discusses a new holistic digital documentation concept – or digital content management paradigm – known in the current scientific literature as *Building Information Modelling*, or BIM, with great potential in the Cultural Heritage and Heritage Conservation Sector (Maietti et al., 2018). The following section introduces BIM implications for existing UNESCO heritage sites (Valero et al., 2018) and recent BIM definition towards standardisation and multi-layer collaboration (Macwilliam & Nunes, 2019). With this newly suggested methodology, different types can now be easily integrated across the entire use space (Pocobelli et al., 2018), thus leveraging feasibility for building scientific knowledge. Therefore, stakeholders can build a solid scientific background without being hindered by cumbersome and rudimentary processes often met in heritage surveys (Maietti et al., 2018).

The chapter begins with a short introduction to BIM and elaborates on its ontology. Next, it demonstrates BIM's competitive strengths concerning speed and efficiency compared to traditional surveys, followed by a short explanation of its continuous growth as an active research area. However, profound differences in complexity, rarity, structural divergence (Idjaton et al., 2021; Pocobelli et al., 2018), and scarcity of data within the Heritage and Conservation undermine BIM extension (Ma et al., 2020). The authors aim at subtle nuances of the BIM concept through a comparison of *Computer-Aided Design* (CAD)–based alternatives, hoping to provide a more solid understanding of this "new" building paradigm shift. Finally, the chapter closes by mentioning key findings of BIM and significant concepts sought in the reviewed literature.

Building Information Modelling (BIM)

BIM is a relatively new process of integrating virtual models of buildings and construction works, along with all the corresponding metadata. This 4D implementation of spatial and time-dimensional digital representation outperforms traditional CAD 3D approaches (Romero-Jarén & Arranz, 2021). Time parameter allows for inter-discipline collaboration among industry sectors (Maietti et al., 2018), with limited unnecessary rework or experts' collisions (Ma et al., 2020). In its basic definition, this standardised and uniform workflow (Croce et al., 2021; López et al., 2017; Pocobelli et al., 2018; Soilán et al., 2020, Axaridou et al., 2014) can be considered a three-step approach, namely: (i) data acquisition, (ii) 3D modelling, and (iii) *Historic Building Information Modelling* (HBIM) object reconstruction. Usually, *BIM for Heritage* – or HBIM – workflow requires a topographic survey with 3D capturing technology, point cloud acquisition, feature extraction, and, finally, primitive fitting and exporting of high complexity organic shapes to a compressed, versatile, and memory-efficient object (Croce et al., 2021; Macwilliam & Nunes, 2019; Pocobelli et al., 2018). This last step consists practically of a cost-effective way of transforming non-modifiable – semantically "poor" and "hard" to interact data – into easily modifiable, semantically enriched, and fully parameterised and reusable "smart" objects (López et al., 2017; Macwilliam & Nunes, 2019; Pocobelli et al., 2018, Mania et al., 2021).

Interoperability is the most fundamental cause of BIM experiencing growth, especially in the *Architecture, Engineering, Construction, and facility Operation* (AECO) field. BIM methodology allows for transparent data collaboration without negotiating on data integrity. Additionally, experts with different backgrounds can contribute simultaneously to the same project with any helpful metric such as dimensions, materials, conditions, and costs (Morbidoni et al., 2020; Romero-Jarén & Arranz, 2021; Soilán et al., 2020). However, this multi-layered and multi-faceted approach, to date, entails high expertise, non-trivial, and quite cumbersome intermittent processes with various tasks being allocated to different model-authoring software (Antón et al., 2018; Pocobelli et al., 2018)

From BIM to HBIM

As stated in the previous paragraph, BIM's success in the AECO industry stems from this new paradigm shift towards multi-functional and multi-faceted collaboration regarding management, monitoring, maintenance, and performance analysis of an existing building throughout its life cycle, from conception to demolition (Croce et al., 2021; Logothetis et al., 2015; López et al., 2017; Pocobelli et al., 2018). Another factor contributing to the expansion of this area – which took place not only in terms of processing results but also concerning the range of its applications – is the growth of interest observed throughout the preservation sector, specifically within the context of the architectural heritage (Bienvenido-Huertas et al., 2020), such as the Horizon 2020 "INCEPTION" European project launch (Maietti et al., 2018). Finally, continuous technological improvements, primarily to image-based 3D rendering, 3D digitising, and machine vision, lead to campaigns' cost reduction and broadening automation alternatives (Croce et al., 2021; Martínez-Carricondo et al., 2020; Pocobelli et al., 2018).

Expansion of this multi-layer BIM integration in contemporary engineering eventually led experts and researchers to explore its possibilities in the demanding realm of archaeology and heritage preservation (Croce et al., 2021; López et al., 2017; Pocobelli et al., 2018). Historically speaking, the original objective of BIM was creating a protocol in order to establish a coherent and reliable library of parametric – or "smart" – objects dedicated to historical typologies. This effort was put forth as a response to the observed complexity of heritage-related exhibits, which poses the need for multidisciplinary approaches from different experts co-work with sometimes conflicting interests (Logothetis et al., 2015; López et al., 2017; Romero-Jarén & Arranz, 2021). Consequently, a new field arose aiming at the creation of the *Heritage BIM* (HBIM), which goes back to 2009 when the first plug-ins were released for commercial BIM platforms, like *ArchiCAD Graphisoft* and *Revit Autodesk*, which by that time had already penetrated the industry workflow (Logothetis et al., 2015; Tkáč et al., 2018).

Researchers and engineers have already tested BIM across many fields and applications, from increasing efficiency and quality, visual data integrity (Khalil & Stravoravdis, 2019) or easing data handling and digesting – since 3D data is more straightforward to the non-expert, if compared to 2D sketches. More advanced applications employ auto-generated disparities maps with high autonomy and cost-effectiveness (Khalil & Stravoravdis, 2019). New possibilities emerged, such as the ability to envisage renovations and restorations with new structural components projected in real-time, virtual replicas of the site (Logothetis et al., 2015). Management-oriented use cases involve energy analysis, multi-thematic sustainability analysis, automatic identification and classification of type and severity of degradation/damages or material deterioration, monitoring, and fast and secure data dissemination under a versatile scheme (Khalil & Stravoravdis, 2019).

Other prominent applications within the context of heritage preservation concern structural simulation, such as *Finite Element Modelling* (FEM) (Antón et al., 2018; Pocobelli et al., 2018), *Computational Fluid Dynamics* (Pocobelli et al., 2018), seismic vulnerability analysis (Antón et al., 2018), virtual reconstructions (Empler et al., 2018; Empler, 2018), and virtual exhibitions utilising *Augmented Reality* (AR) (Baik, 2021), GIS and BIM data integration (Pocobelli et al., 2021), and synthetic point cloud data generation (Babacan et al., 2017; Ma et al., 2020; Pierdicca et al., 2020).

All above-mentioned possibilities contributed to the converging behaviour between heritage and BIM, which can now be referenced as a standard workframe for strategic planning and decision-making policies (Logothetis et al., 2015), whilst harvesting collaboration between experts and stakeholders (Bienvenido-Huertas et al., 2020; Khalil & Stravoravdis, 2019; Logothetis et al., 2015; Pocobelli et al., 2018) in a multi-faceted and holistic approach (Croce et al., 2021; Logothetis et al., 2015; Maietti et al., 2018; Soilán et al., 2020).

HBIM-Inhered Obstacles to Standardisation

Nevertheless, heritage conservation–BIM integration poses some serious challenges, not only because of artefact high complexity – which diverts from the standard architectural "grammar" – but also due to its susceptibility to error propagation, which undermines standardisation scheme (Croce et al., 2021; Garagnani & Manferdini, 2013; Pocobelli et al., 2018). In contrast to the vast information exhibited in new and under-construction buildings, the field of heritage conservation is not uncommon to fall into vaguely (if not at all undocumented) documented historic data, for example, restoration efforts or sparsely distributed records concerning geometry and/or construction or renovation processes (Khodeir et al., 2016; León-Robles et al., 2019).

Even though BIM offers quite interesting opportunities in the field of heritage conservation (Antón et al., 2018; Borin & Cavazzini, 2019; La Russa & Santagati, 2020; Logothetis et al., 2015; López et al., 2017; Martínez-Carricondo et al., 2020; Pocobelli et al., 2018), it also encompasses a hypothesis-related opposition. As Pocobelli et al. (2018) emphasise about BIM for heritage, historical assets intentionally tend to "disobey" uniformity rules which are, by contrast, implicitly assumed in BIM. On the other hand, BIM efficiency is built upon the idea of producing "standardised" serialised components that can easily be numerically "constrained" and therefore imported to and retrieved from dedicated open-access libraries. Therefore cost-effectiveness of the BIM principle stems from – and derives upon – re-usability "hypothesis" which in principle is violated within the Heritage sector, which is mostly occupied by hand-crafted, unique, and complex structures (Martínez-Carricondo et al., 2020; Morbidoni et al., 2020; Pocobelli et al., 2018). Conversely, automating and producing BIM-compliant containers for those refined and hard-to-produce heritage artefacts is an added function of the Conservation field (Croce et al., 2021).

3D Modelling for HBIM

Modelling of the existing infrastructure incorporates a labour-intensive and time-consuming reverse engineering process which, in BIM glossary, primarily falls under the definition of *Scan-to-BIM* (Croce et al., 2021; Pocobelli et al., 2018; Valero et al., 2018). *Scan-to-BIM* indeed is composed of three distinct phases: (i) "data collection", (ii) "segmentation", and (iii) "HBIM reconstruction" (Croce et al., 2021). Pocobelli et al. (2018), referring to HBIM maturity – and specifically to the ability of this multi-purpose scheme to handle real-time data visualisation, pose the need to further divide this last one step into two different procedures. The first one refers to the HBIM reconstruction during which parametric objects are formed. The second one corresponds to metadata embedding and integration upon these previously created objects. In the forthcoming part of this section, each of the four in total aforementioned steps will be briefly explored only to serve the need within an introductory chapter.

Data Acquisition

3D surveying nowadays sees a radical improvement due to recent advances in active and passive sensors, such as *Terrestrial LASER Scanners* (TLS) or image-based 3D representation technology utilising air-borne or ground means (Braun & Borrmann, 2019; Idjaton et al., 2021; Macwilliam

& Nunes, 2019; Valero et al., 2018, Maravelakis et al., 2014a). These high-recording speed-capturing devices used in modern 3D surveys provide for increased accuracy and high-fidelity data, either independently or through data fusion and overlapping (Croce et al., 2021). Currently, widely exploited active elements are TLS, such as *Leica ScanStation C10 3D Laser Scanner* – transmitting a laser pulse beam across a 360-sphere inscribed by two concurrently rotating angles or plates, revolving around azimuth and declination levels (Antón et al., 2018). On the other end of the spectrum, passive elements differ from their active counterparts – that is, they exploit already emitted and multi-scattered electromagnetic radiation in order to register their surroundings in a local point cloud (Pocobelli et al., 2018). Even though these passive means tend to underperform with respect to the ones actively engaged in final outcome quality (Braun & Borrmann, 2019), their popularity increases thanks to their relatively low price (Bienvenido-Huertas et al., 2020; Croce et al., 2021). One of the most characteristic representatives, heavily exposed to Scan-to-BIM applications, is digital photo-cameras that have their photos processed through a 2D photogrammetry software able to resolve camera positions and replicate 3D geometry by *Structure-from-Motion* (SfM) algorithms (Antón et al., 2018; Croce et al., 2021; Pocobelli et al., 2018).

Terrestrial-Generated Data

TLS outperforms every other available 3D real scene–capturing device, in terms of dense point cloud registration efficiency. Indeed, they operate at a pace of approximately one million points per second – without limitations being imposed by external factors such as ambient illumination or other environmental conditions. At the same time, large construction or monumental sites can now be mapped remotely, with non-invasive methods including hazardous or inaccessible and hard-to-reach locations on site. Data recording speed varies depending on the following: (i) user set parameters, for example, resolution or scanning with colour; (ii) project prerequisites, for example, quality, accuracy, and reliability; and (iii) site complexity or decorative elements. Speed may fluctuate from about 2 min to 10 min and 9 min to 20 min for indoor and outdoor applications, respectively (Tkáč et al., 2018), with a perceivable accuracy of approximately 5 mm (Antón et al., 2018). The only caveat associated with TLS technology is the huge amount of data typically produced within 3D survey (Antón et al., 2018; Morbidoni et al., 2020). On this matter, Morbidoni et al. (2020) reported on their survey a sum of 78 different scans, which yielded back 1.2 billion points at an estimated accuracy of about 1 cm at 100 m distance range.

TLS, with its fast and accurate production scheme, has revolutionised BIM integration workflow. Additionally, it has allowed for two different disciplines to be developed within the building industry, namely *Scan-to-BIM* and *Scan-vs-BIM* – both with serious implications in engineering practices (Antón et al., 2018; Valero et al., 2018). *Scan-to-BIM* originates from the necessity to document an already existing reality scene, with its geometry deviating from the theoretical architectural drawing, and aiming at establishing a BIM framework for future proofing and processing. This "as-is" or "as-built" digital documentation of the existing building features falls within the notion of reverse engineering. Conversely, the *Scan-vs-BIM* concept introduces a new paradigm in technology integration, with the primary objective of closely monitoring in real time – using on-site observations and at millimetre range accuracy – deviations between implemented and designed construction works. These are two of the most characteristic examples of added value to the space of relating on-side "as-built" data to virtual "as-designed" in near real-time context (Tkáč et al., 2018).

UAV Generated Data

Air-borne technology is also gaining popularity thanks to new emerging technology, especially in the field of *Unmanned Aerial Vehicles* (UAVs). UAVs can carry *Light Detection and Ranging* (LiDAR) payload with high-precision *Global Positioning System* (GPS), achieving a total geo-reference accuracy of no more than a couple of centimetres – which coincides with the common requirements met in the field of BIM (Martínez-Carricondo et al., 2020). Moreover, both grounded and aerial means can be deployed with additional sensors, such as on-board, thermal, or optical scanners, resulting in high redundancy, diversification, and densification of the 3D sampled space (Antón et al., 2018; Croce et al., 2021).

Aerial surveying, even though most frequently used in conjunction with terrestrial inspection (Braun et al., 2020), exhibits pros and cons, when compared to TLS or other ground-based alternatives, briefly explored herein. Starting from the former, UAVs are immune to the physical dimensions of the property being examined and can operate autonomously at a very fast pace – only being limited by the camera's shutter-speed inferred distortions, and this only applies to image-based reconstruction utilising SfM algorithms (Martínez-Carricondo et al., 2020). Additionally, at this high speed, high spatial fidelity results in significant cost reduction, which could reach as high as 40% of that conventionally expected (Isailović et al., 2020). Finally, it is worth mentioning that UAV-borne observations, given their manufacturers' specifications, can reach spatial accuracies in the order of plus or minus 15 mm in both vertical and horizontal directions – which equals accuracy perceived by ground means. However, sub-centimetre level of accuracy in the positional arguments requires some extra steps to be taken in the photogrammetric workflow, that is, establishing *Ground Control Points* (GCP), or GPS signal restitution by implementation of *Precise Positioning Kinematic* solutions to it (Martínez-Carricondo et al., 2020).

Compared to conventional terrestrial means, UAVs and their ground control units in charge of navigation, such as *Pix4D and DroneDeploy* (Braun et al., 2020), rely on GPS signals for path planning and executing. Therefore, UAVs cannot operate, or fail in autonomous operation inside buildings or near cluttered locations, that is, where the GPS signal attenuates or is even lost (Braun et al., 2020; Martínez-Carricondo et al., 2020).

Regarding data acquisition speed and efficiency, photogrammetry SfM-based reconstruction relies on CPU- and GPU-demanding tasks to produce depth maps for the reconstructed scenes. Therefore, image-based 3D modelling usually requires time-and-resource-intensive processes, due to the significant number of photos necessary in heritage applications, pushing the limits of the currently available hardware and software technology. For example, Martínez-Carricondo et al. (2020) report professionals working for 15 h to finalise an SfM-based "clean" 3D model.

Furthermore, except for time and processing power issues, photogrammetry presents weaknesses, either in the case of uniform or texture-less structures that present visual homogeneity or in the case of high transmissibility surfaces, such as windows (Braun et al., 2020). Last but not least, UAV expansion is now decelerating due to increased bureaucracy and UAV operation costs after new local and international provisions (Isailović et al., 2020).

Point Cloud Processing and Segmentation

Segmentation prerequisites an already aligned, collimated, clean, and clutter-relieved point cloud (Ma et al., 2020). Specifically, 3D data from one or multiple sources, that is, instruments, are referenced to a common frame – a process known as point cloud registration – with different datasets, from multiple sources, and multiple viewing angles and measuring locations are projected on a ground truth Cartesian coordinate system (Braun & Borrmann, 2019; Garagnani & Manferdini, 2013; Pocobelli et al., 2018). In some cases, preliminary steps are required, before the actual campaign, to eliminate the possibility of dense point cloud coverage being ruined by shadow or hidden unseen, from the sensor, areas (López et al., 2017).

In case of increased positional accuracy requirements, point cloud registration is followed by a transformation of the internal coordinate XYZ frame to a local or universal coordinate system. This process of geo-tagging requires user-set, carefully selected, ground truth observable targets, a.k.a. GCPs, that are meant to be used as a pivoting plane, having their spatial accuracy restituted through GPS post-processing algorithms (Antón et al., 2018). Once the point cloud is merged and registered, each XYZ point in the *Euclidean Space* – along with any other available information attributed to that specific location, such as backscattering coefficient or colour – needs to be classified with a proper architectural element-wise class tag (Morbidoni et al., 2020). The latter attracts most of the scientific research since it is the most rigorous, labour-, time-, resources-, and expertise-intensive part of the HBIM workflow – and usually met under the definition of "semantics" or "semantic segmentation". In general, this process addresses all the available heuristics of converting a rich, high-density, unstructured point cloud to a set of fully constrained and tabulated BIM readable "smart" elements, whilst repeating this loop for each one of the individual components within each sub-space of the entire asset (Antón et al., 2018; Croce et al., 2021; Morbidoni et al., 2020).

The point cloud segmentation step, as explored in the previous paragraph, entails a recursive iteration where meaningful geometries, or features, need to be extracted from the heritage point cloud (Bassier & Vergauwen, 2020; Croce et al., 2021). Once this process resolves, each row of the point cloud data file – corresponding to a position in the 3D space – will have an element-wise class descriptor allowing for point cloud division or segmentation (Croce et al., 2021). For the heritage conservation field, and depending on the *Level of Detail* set (Pocobelli et al., 2018; Valero et al., 2018), the descriptor's class labels may be capitals, vaults, windows – or, in view of modern structures, steel beams, concrete slabs, and ducts (Logothetis et al., 2015) – or, if simpler geometries are traced, vertical planes, walls, floors, and ceilings (Croce et al., 2021).

Mesh segmentation ranges from simple, manual, and visual-based geometry extraction techniques – where experts trace down exhibited geometry aimed by CAD's "smart" feature capturing capabilities (Garagnani & Manferdini, 2013) – to semi-automatic ones, where either segmentation or HBIM reconstruction is machine-intercepted (Antón et al., 2018; Soilán et al., 2020) up to highly sophisticated and fully automated procedures utilising supervised (Valero et al., 2018) or unsupervised *Machine Learning* (ML) algorithms (Ma et al., 2020), artificial intelligence (AI) (Bienvenido-Huertas et al., 2020), or deep learning algorithmic schemes (Idjaton et al., 2021; Ma et al., 2020; Morbidoni et al., 2020).

In automation endeavours, the main branches are model-driven or data-driven methods. In the first case, a raw unstructured point cloud is segregated based on its congruence to a predefined object, shape, or identity, as measured by fitting coefficients. Quite popular approximation of this approach implements *RANSAC* (Ma et al., 2020; Valero et al., 2018), or Region growing algorithms – where the former tries to match foreseen geometry to a prescribed geometry, denoted by a model, and the latter attempts to extend seeded region based on similarity or dissimilarity with its adjacent area. These model-driven algorithms fail in robustness, re-usability, and generalisability. On the other hand, data-driven algorithms try to predict objects, either by feature mapping or in an unsupervised manner. The latter approach, which usually requires deep learning 3D-enabled networks, such as *PointNet or Dynamic Graph Convolutional Neutral Network*, outperforms other methods in terms of robustness and generalisability – however, it requires a large number of training samples which is hard to get (Ma et al., 2020).

Broadly speaking, BIM workflow automation starts with a high complexity large-scale buildings or monuments and can reach up to stone-by-stone segmentation (Idjaton et al., 2021; Valero et al., 2018). Specifically, this processing cycle entails a repetitive loop of finely tuned algorithms, which run recursively on a (coloured) point cloud trying to maximise their performance, in either supervised or unsupervised manner. Once predictability reaches an acceptable threshold, the training phase ends and models re-runs on the rest of the "unseen" dataset; afterwards, it predicts and assigns class-or-feature labels on each one of the individual points, back-propagating and transferring a priori "knowledge", gained during the training session, on the entire dataset Croce et al., 2021). Segmentation is still a subject of research and open debate with different authors, adopting different ML heuristics, supervised (Bienvenido-Huertas et al., 2020), unsupervised, or deep learning (Koo et al., 2021; Morbidoni et al., 2020) – in a co-effort to speed up and error-proof digital documentation of architectural assets through robustness and easiness of implementation by the non-expert user (Croce et al., 2021; La Russa & Santagati, 2020; Logothetis et al., 2015; Pocobelli et al., 2018; Valero et al., 2018).

HBIM Applications

HBIM was developed by Murphy and Pavia in 2009 (Murphy, McGovern and Pavia, 2009). Since then, its usage has sharply increased in different applications (Pocobelli et al., 2018). Specific parametric libraries have been created, such as the *Jeddah Historic Building Information Modelling* (JHBIM), to fit and represent local architectural style and construction technique; endeavours have been made to integrate BIM models with GIS, either for localisation purposes or for condition mapping; monitoring has been attempted, either using *Internet of Things* (IoT) technologies or depicting condition and environmental behaviour – just to cite a few main topics.

This section will explore a plethora of topics where the application of HBIM has been of support. Please note this list does not mean to be exhaustive – the aim here is to give an idea of the variety of fields where BIM has been used during the last decades.

HBIM Application for Conservation and Restoration Purposes

Conservation and, consequently, restoration are hot topics for the built heritage field. Historic England considers it pivotal to understand, maintain, and protect for future generations our built heritage for different reasons, including understanding our culture and history, building our community, educating our young generations, and producing economic revenue thanks to tourism (Historic England, 2008; Pocobelli et al., 2021). BIM is a tool that can be used to support the understanding and protection of our built heritage, specifically concerning information collection (Pocobelli et al., 2018). Conservation and restoration of existing buildings through BIM is a topic which has been well studied; in this section, we will explore two main applications of BIM for conservation/restoration purposes:

1. *Historic Building Information Modelling* (HBIM)
2. *Jeddah Historic Building Information Modelling* (JHBIM)

HBIM and JHBIM are both parametric libraries designed for a specific architectural style. HBIM has been developed by Murphy et al. (2009) to represent classical European and Irish architecture of the seventeenth and eighteenth centuries (Murphy et al., 2013), whilst JHBIM is a variation of HBIM intended to portray typical Old Jeddah architecture (Baik et al., 2014; Baik et al., 2013; Baik et al., 2015). Parametric libraries are a collection of predefined building elements that can be adapted in size (and in other predefined parameters) whenever needed (Fai & Sydor, 2013; Murphy et al., 2013; Oreni et al., 2013; Pocobelli et al., 2018; Saygi et al., 2013; Worrel, 2015).

HBIM can be intended as a form of procedural modelling, taking shape from shape grammars, as introduced by Stiny and Gips during the 1970s (Stiny & Gips, 1972). In this sense, shape grammars can be intended as a "basic vocabulary" that will build following specific "production rules" (Pocobelli et al., 2018). Murphy et al. (Chenaux et al., 2011; Dore et al., 2015; Dore & Murphy, 2012; Dore & Murphy, 2013a; Dore & Murphy, 2013b; Dore & Murphy, 2014; Dore & Murphy, 2017; Murphy, 2012; Murphy et al., 2021; Murphy et al., 2009; Murphy et al., 2013) were the first to use this notion to produce a complete architectural style for classical European architectural.

As previously stated, JHBIM builds on HBIM fundaments. JHBIM can be intended as a specific HBIM parametric library specific for the Old Jeddah architectural style (Baik et al., 2013, 2014, 2015). Additionally, JHBIM is also thought to be capable of displaying different historical layers, local traditional construction techniques, and so on (Baik et al., 2013, 2014, 2015).

HBIM Applications on Management Requirements

A few attempts have been made to use HBIM technology within the management field. Specifically, two main subfields can be identified:

- *Facility Management* (FM)
- Behavioural predictions

Concerning the FM field, McArthur et al. (2018) propose an integration between the BIM model and ML algorithms to flag and track maintenance issues. Specifically, they create an innovative way to collect system failure and maintenance information through *Work Orders*; these reports are sent online through real-time ML algorithms and can be dealt with immediately by the FM team. Thanks to the building virtualisation provided by BIM, the FM team can immediately spot the location and the gravity of reported issues. Additionally, this method classifies reported issues and enables live feedback with follow-up questions to reporting users.

On the same matter, Pour Rahimian et al. (2020) suggest a hybrid method to facilitate FM operations, which consists in enabling easier access to high-volume data produced by HBIM models regarding maintenance processes. Indeed, during its lifetime, a building produces a massive amount of data which proceeds from inspections, repairs, and so on (Pour Rahimian et al., 2020). Dealing with such load can be extremely complicated and time-consuming (Arias et al., 2007; Pocobelli et al., 2018). To solve this issue, Pour Rahimian et al. (2020) propose a virtual reconstruction with integration of BIM technology and videogame *Virtual Reality* (VR) environment. The authors suggest that their platform can be used both during the construction phase – to follow progress and present results to clients – and afterwards – to support maintenance.

In the matter of behavioural predictions, Simeone's work (Simeone, 2018) highlights the possibility of using BIM on existing buildings to enable predictions of both users' behaviour and condition, through a specific methodology called *Event-Based Modelling.* Indeed, Simeone creates a virtual environment of an existing building and predicts its possible usage and occupancy. This information can be later used to assess whether it is possible to re-use the building, and how this usage could affect its condition.

HBIM Applications for Monitoring

HBIM has started to be used for monitoring purposes very recently. Indeed, until 2018, there were very few examples of its usage in this field (Pocobelli et al., 2018).

Environmental monitoring and simulations studies are very limited (Pocobelli et al., 2021); examples include Seghier et al.'s (2017) work, as well as our own work (Pocobelli et al., 2021). They both rely on Dynamo to display environmental modelling. Seghier et al. (2017) produce a BIM model and link it to a Dynamo script to create a workflow for *Green Building Designing.*

Regarding condition representation, HBIM was started to be used only recently (Pocobelli et al., 2021). Lopez-Gonzalez et al. (2016) produce a virtual 3D model based on GIS, where geo-referencing coordinates are used to both localise condition – in this specific case, moisture – and to depict its development over time as well. Another endeavour has been made by the Politecnico di Milano team, where they were able to insert into the HBIM model degradation patterns (Oreni et al., 2013). A similar result was produced by Saygi et al. (2013) using GIS.

Whilst attempts to depict conditions and to represent environmental modelling are very limited, there is a plethora of studies regarding the IoT (Pocobelli et al., 2021); nonetheless, as it proceeds from Tang et al. (2019), studies concerning integration between IoT and HBIM are still in their conceptual form, and theoretical frameworks are provided without a practical solution yet. Specifically, Tang et al. (2019) identify five different integration methods:

1. *BIM APIs with database.* This method has been tried by Marzouk and Abdelaty (2014), where they integrate a virtual model with environmental data through *Microsoft ACCESS.*
2. *BIM to be transformed into a relational database using new schemes.* So far there haven't been any attempts to our knowledge (Pocobelli et al., 2021).
3. *Creation of a new query language.* This approach has been attempted by Mazairac and Beetz (2013).
4. *Using the semantic web.* Curry et al. (2013) have investigated this method, which seemed reliable for data storage and sharing (Pocobelli et al., 2021).
5. *Hybrid.* This technique is a mixture between methods (1) and (3) and has been experimented by Hu et al. (2016), who create a mixed system with building performance data and building context data.

HBIM Applications on Structural Simulations

Several attempts on integrating HBIM with structural simulations exist. Since the very beginning, HBIM has been extensively used to analyse and reproduce structural conditions of heritage (Motsa et al., 2020; Pocobelli et al., 2018; Tapkin et al., 2022). Whilst several examples exist, we will focus here on a couple of studies to show the state of the art.

Marzouk et al. (2020) develop an HBIM model from *LiDAR* data, and they perform a structural simulation using FEM analysis. Their results show great detail on the structural condition of the case study, and these results are extremely precious for developing scientifically informed recommendations to stakeholders.

On the same matter, Barazzetti et al. (2015) develop a method in two steps, where an HBIM model is produced, and it is afterwards analysed through FEM methods. Additionally, the authors have rationalised building geometry, keeping irregularities, though, to better facilitate FEM analysis.

HBIM Applications for Virtual Reconstructions

Virtual reconstructions can be performed for several reasons. In this section we will explore four different kinds of virtual reconstructions:

1. For earthquake protection, through the *Assessment Reconstruction Information Modelling* (ARIM) methodology
2. For maintenance and live analysis, through *Forge/Dasher*
3. For sharing purposes through cloud technology and websites (*Tridify, Sketchfab*, and *Arch3D*)
4. For holistic representation through the *Digital Twin*

Due to the 2016 earthquake in Central Italy, the town of Amatrice completely collapsed. This episode is a reminder of seismic vulnerability, as well as of the need to protect both humans and heritage (Empler, Calvano, and Caldarone, 2018). Empler's team (Empler, 2018; Empler et al., 2018) developed a specific BIM procedure for earthquake disasters, called ARIM, the aim of which is to scientifically document the condition of seismically affected buildings to endorse seismic retrofit. Due to the always-changing nature of information in earthquake-affected areas, data representation is an issue, since data obsolescence can happen at any time. The authors (Empler, 2018; Empler et al., 2018) manage to solve this issue using a *data fusion* framework. The resulting virtual model is a "responsive" model, which is automatically updated anytime new information is available. The model can be accessed, and it provides specific information on façade deformations, collapse mechanisms, and so on.

Concerning the second type of virtual reconstruction, Autodesk has developed a cloud-based platform called *Forge*, where users can upload their BIM model and perform any type of live analysis (Autodesk, n.d.). A demo version of *Forge*, called *Project Dasher 360*, has been created by Autodesk researchers, in which it is possible to monitor the live environmental parameters of the digitised building (Autodesk, 2018). The BIM model uploaded to *Dasher* is connected to physical sensors in the real building. These sensors can be clicked by users, and they are able to produce plots (Pocobelli et al., 2021).

Model sharing can be successfully obtained through uploading to dedicated websites. There is a plethora of platforms that can be used, either for commercial use or for research purposes. *Sketchfab* (Sketchfab, n.d.) is a commonly used viewer that provides uploading/downloading services for 3D models (Pocobelli et al., 2021). A more BIM-oriented alternative is *Tridify*, which is an online viewer that supports data attachment, as well as common 3D viewing operations, such as measuring (Tridify, n.d.). Finally, in the academic field, there have been attempts to produce heritage-focused BIM platforms, such as *Arch3D* – now deprecated (Pocobelli et al., 2021) and web-based point cloud viewers (Maravelakis et al., 2014b).

Finally, virtual representations can be achieved through the *Digital Twin* (DT). The DT conceptually exists since the early 2000s (Boje et al., 2020), and research is very active in this field, especially because of its IoT and AI components. Indeed, Boje et al. (2020) define the DT as a holistic framework with a semantically enhanced BIM, which is boosted with IoT sensors and AI technology.

HBIM Integration with GIS

Integration between GIS and BIM is a very popular topic in the heritage sector since several studies have been carried out (Pocobelli et al., 2021). GIS can be intended as a 3D tool with geographical coordinates, where every object can be geo-referenced (Dore & Murphy, 2012). It is commonly used for large-scale projects, such as landscape studies and archaeological site representations. However, its geo-referencing nature makes it particularly suitable for condition mapping in heritage buildings (Rinaudo et al., 2007).

Several studies exist on attempting HBIM integration with GIS technology. Examples include the work conducted by Bruno et al. (2020). Authors develop an HBIM-GIS integration through a dedicated online platform called *Chimera*, where different historic layers can be restituted on demand (Bruno et al., 2020). Similarly, Albourae et al. (2017) use a GIS database to store data that cannot be directly embedded into the BIM model, such as non-architectural information and geographic data. This list is non-exhaustive, according to authors' perspectives and, however, aims at providing a good insight into the current state-of-the-art endeavours in BIM extensions.

Hybrid BIM

The term *hybrid* is extremely vague and could refer to any attempt to integrate BIM with any other technology. In this section, we will consider as *Hybrid BIM* all these efforts where BIM has been integrated with point cloud data or similar.

Literature shows that there have been attempts to produce hybrid BIM models, for disparate purposes. For instance, Kim et al. (2020) carry out a study where they compare two different models: the BIM one and a second one produced by point cloud data. Through this comparison, they can effectively assess the accuracy of the above-mentioned virtual representations; additionally, they use this information to efficaciously track construction progress for new buildings (Kim et al., 2020). Similarly, Amano et al. (2019) produce a mixed model with 3D point cloud data, hyperspectral data, and BIM building elements. Authors observe that since 3D point cloud data is not semantically significant, this data should be referred to external sources, in order to be used in BIM (Amano et al., 2019). In this study, the authors aim at supporting the retrofit of existing buildings to reduce carbon emissions (Amano et al., 2019).

HBIM Applications for Promotion and Leverage

Digitisation of existing buildings could potentially be a very clever solution to endorse tourism (Baik, 2021), or even to provide virtual visits during events similar to the COVID-19 epidemic.

A few studies exist regarding this topic. Baik (2021) uses HBIM integrated with VR and AR to create a virtual environment for the *Zainal Historical House* in Old Jeddah; with this solution, users can access the VR environment and experience in-depth knowledge. The author also highlights that this kind of solution is beneficial to both experts and visitors since HBIM models could be used as a promotional and educational material (Baik, 2021). On the same matter, Barazzetti and Banfi (2017) highlight the importance of BIM usage on mobile devices, specifically for professional purposes. Indeed, this technique enables simultaneity and time efficiency (Barazzetti & Banfi, 2017). Additionally, professionals can be on-site more often, having with them their BIM model, and having the possibility to be connected with other specialists on the same model (Barazzetti & Banfi, 2017). This technique uses an integrated BIM with AR and VR, depending on need, where the BIM model is uploaded to a cloud and therefore accessible to all stakeholders (Barazzetti & Banfi, 2017). Authors also provide examples of the usage of mixed BIM with AR and VR for touristic applications, such as the case study of Castel Masegra, Italy (Barazzetti & Banfi, 2017).

HBIM for Modelling Archaeological Uncertainty

The unprecedented increase of new technological affordances within the frame of sustainability and heritage conservation, such as VR and AR (Barazzetti & Banfi, 2017), has raised scientific concerns regarding visual representation accuracy. Specifically, scientists posit that the optical reconstruction of many sites may lack precision in favour of aesthetics and pleasingness of the audience. Therefore, new scales of conformity need to be enacted, such as the Archaeological Uncertainty, which – combined with BIM and new 3D surveying technology – may constrain arbitrariness and bias (Mania et al., 2021; Sifniotis et al., 2007; Sifniotis et al., 2010).

Archaeological uncertainty now deviates from the classical variance-like definition aiming to maximise different experts' agreement on the final product – whilst utilising advanced probabilistic functions and fuzzy logic to rationalise the gap between excavation and realisation. BIM integration may subjectively transform that knowledge gap into an alternatively built scenario and resend it for evaluation from the experts. Mania et al. (2021) applied this technique, reflecting on different configurations of the decorative metopes excavated in the area of the Athenian Treasury at Delphi under the project 3D4DEPLHI (https://3d4delphi.gr/en/archaeological-uncertainty/) funded by both national and European expenses.

Summing up, in this section a few examples of HBIM applications were explored seeking to provide a robust background regarding HBIM fundamentals.

Discussion

In conclusion, BIM is a powerful tool able to support the understanding and protection of worldwide built heritage, specifically through its data collection capacity with a substantial impact on education, sustainable tourism, and economic growth. Successful applications of BIM fall within the scope of conservation and restoration (Historic England, 2008; Pocobelli et al., 2021), management (Arias et al., 2007; McArthur et al., 2018; Pocobelli et al., 2018; Simeone, 2018), monitoring (Lopez-Gonzalez et al., 2016; Pocobelli et al., 2018; Seghier et al., 2017), data visualisation (Lopez-Gonzalez et al., 2016), systems integration (Bruno et al., 2020; Kim et al., 2020), data enrichment (Amano et al., 2019; Croce et al., 2021; Logothetis et al., 2015), decentralisation (Croce et al., 2021; Logothetis et al., 2015), and sustainable growth (Baik, 2021; Barazzetti & Banfi, 2017).

Active research now engages the time-consuming and labour-intensive processes of the "as-is" or "as-built" BIM implementation, mainly focusing on mesh-segmentation automation (Maietti et al., 2018). Endeavours cover a wide span starting from smart-feature recognition manual-driven methods (Garagnani & Manferdini, 2013) to highly sophisticated complete automated ones utilising ML algorithms (Ma et al., 2020; Valero et al., 2018), deep learning (Idjaton et al., 2021; Ma et al., 2020; Morbidoni et al., 2020), and AI (Bienvenido-Huertas et al., 2020).

Scientists seek to discover new and more robust heuristics for mesh segmentation and primitives extraction (Macwilliam & Nunes, 2019; Pocobelli et al., 2018), with new parametric libraries created as we speak for specific architectural typologies (Murphy et al., 2009). JHBIM consists of a successful variation of HBIM intended to depict Old Jeddah architectural monuments through a collection of fully defined building elements (Baik et al., 2013). Alternatives regarding primitive extraction are also met in the current literature with Dynamo script suggested for that exact reason as an alternative workflow for *Green Building Designing* (Seghier et al., 2017).

Finally, ML algorithms are introduced in all stages of HBIM. From autonomous maintenance and tracking of building status to real-time online information regarding the time and severity of future works. Consequently, virtualisation affordances of the BIM technology can alleviate the localisation and gravity of the reported issues (Arias et al., 2007; Pocobelli et al., 2018; Pour Rahimian et al., 2020) to mesh segmentation and semantisation (Croce et al., 2021; La Russa & Santagati, 2020; Logothetis et al., 2015; Pocobelli et al., 2018; Valero et al., 2018).

Acknowledgements

This research forms part of the project 3D4DEPLHI, co-financed by the European Union and Greek funds through the Operational Program Competitiveness, Entrepreneurship and Innovation, under the call "Specific Actions, Open Innovation for Culture" (project code: T6YBΠ-00190).

References

Albourae, A.T., C. Armenakis, and M. Kyan. 2017. "Architectural Heritage Visualization Using Interactive Technologies." *International Archives of the Photogrammetry, Remote Sensing and Spatial Information Sciences – ISPRS Archives* 42 (2W5): 7–13. https://doi.org/10.5194/isprs-archives-XLII-2-W5-7-2017.

Amano, Kinjiro, Eric C.W. Lou, and Rodger Edwards. 2019. "Integration of Point Cloud Data and Hyperspectral Imaging as a Data Gathering Methodology for Refurbishment Projects Using Building Information Modelling (BIM)." *Journal of Facilities Management* 17 (1): 57–75. https://doi.org/10.1108/JFM-11-2017-0064.

Antón, D., B. Medjdoub, R. Shrahily, and J. Moyano. 2018. "Accuracy Evaluation of the Semi-Automatic 3D Modeling for Historical Building Information Models." *International Journal of Architectural Heritage* 12 (5): 790–805. https://doi.org/10.1080/15583058.2017.1415391.

Arias, P., J. Armesto, D. Di-Capua, R. González-Drigo, H. Lorenzo, and V. Pérez-Gracia. 2007. "Digital Photogrammetry, GPR and Computational Analysis of Structural Damages in a Mediaeval Bridge." *Engineering Failure Analysis* 14 (8): 1444–57. https://doi.org/10.1016/j.engfailanal.2007.02.001.

Autodesk. n.d. "Autodesk Forge." Accessed June 6, 2019. https://forge.autodesk.com/.

Axaridou, A., I. Chrysakis, C. Georgis, M. Theodoridou, M. Doerr, A. Konstantaras, and E. Maravelakis. 2014. "3D-SYSTEK: Recording and Exploiting the Production Workflow of 3D-Models in Cultural Heritage." In *IISA 2014, The 5th International Conference on Information, Intelligence, Systems and Applications*, 51–6. https://doi.org/10.1109/IISA.2014.6878745.

Babacan, K., L. Chen, and G. Sohn. 2017. "Semantic Segmentation of Indoor Point Clouds Using Convolutional Neural Network." *ISPRS Annals of the Photogrammetry, Remote Sensing and Spatial Information Sciences* 4 (4W4): 101–108. https://doi.org/10.5194/isprs-annals-IV-4-W4-101-2017.

Baik, Ahmad. 2021. "The Use of Interactive Virtual BIM to Boost Virtual Tourism in Heritage Sites, Historic Jeddah." *ISPRS International Journal of Geo-Information* 10 (9): 577. https://doi.org/10.3390/ijgi10090577.

Baik, A., J. Boehm, and S. Robson. 2013. "Jeddah Historical Building Information Modeling 'JHBIM' Old Jeddah – Saudi Arabia." *International Archives of the Photogrammetry, Remote Sensing and Spatial Information Sciences – ISPRS Archives* 40 (5W2): 73–78. https://doi.org/10.5194/isprsarchives-XL-5-W2-73-2013.

Baik, A., A. Alitany, J. Boehm, and S. Robson. 2014. "Jeddah Historical Building Information Modelling 'JHBIM' Object Library." In *ISPRS Annals of the Photogrammetry, Remote Sensing and Spatial Information Sciences, Volume II-5, 2014 ISPRS Technical Commission V Symposium*, 23–25 June, Riva Del Garda, Italy II–5 (June), 41–7. https://doi.org/10.5194/isprsannals-II-5-41-2014.

Baik, A., R. Yaagoubi, and J. Boehm. 2015. "Integration of Jeddah Historical BIM and 3D GIS for Documentation and Restoration of Historical Monument." *ISPRS – International Archives of the Photogrammetry, Remote Sensing and Spatial Information Sciences* XL-5/W7 (5W7): 29–34. https://doi.org/10.5194/isprsarchives-XL-5-W7-29-2015.

Barazzetti, Luigi, and Fabrizio Banfi. 2017. "Historic BIM for Mobile VR/AR Applications." In *Mixed Reality and Gamification for Cultural Heritage*, edited by Marinos Ioannides, Nadia Magnenat-Thalmann, and George Papagiannakis. Cham: Springer International Publishing. https://doi.org/10.1007/978-3-319-49607-8.

Barazzetti, Luigi, Fabrizio Banfi, Raffaella Brumana, Gaia Gusmeroli, Mattia Previtali, and Giuseppe Schiantarelli. 2015. "Cloud-to-BIM-to-FEM: Structural Simulation with Accurate Historic BIM from Laser Scans." *Simulation Modelling Practice and Theory* 57 (September): 71–87. https://doi.org/10.1016/j.simpat.2015.06.004.

Bassier, M., and M. Vergauwen. 2020. "Unsupervised reconstruction of Building Information Modeling Wall Objects from Point Cloud Data." *Automation in Construction* 120 (November 2019): 103338. https://doi.org/10.1016/j.autcon.2020.103338.

Bienvenido-Huertas, D., J.E. Nieto-Julián, J.J. Moyano, J.M. Macías-Bernal, and J. Castro. 2020. "Implementing Artificial Intelligence in H-BIM Using the J48 Algorithm to Manage Historic Buildings." *International Journal of Architectural Heritage* 14 (8): 1148–1160. https://doi.org/10.1080/15583058.2019.1589602.

Boje, Calin, Annie Guerriero, Sylvain Kubicki, and Yacine Rezgui. 2020. "Towards a Semantic Construction Digital Twin: Directions for Future Research." *Automation in Construction* 114 (January): 103179. https://doi.org/10.1016/j.autcon.2020.103179.

Borin, P., and F. Cavazzini. 2019. "Condition Assessment of RC Bridges. Integrating Machine Learning, Photogrammetry and BIM." *International Archives of the Photogrammetry, Remote Sensing and*

Spatial Information Sciences – ISPRS Archives 42 (2/W15): 201–208. https://doi.org/10.5194/isprs-archives-XLII-2-W15-201-2019.

Braun, A., and A. Borrmann. 2019. "Combining Inverse Photogrammetry and BIM for Automated Labeling of Construction Site Images for Machine Learning." *Automation in Construction* 106 (June): 102879. https://doi.org/10.1016/j.autcon.2019.102879.

Braun, A., S. Tuttas, A. Borrmann, and U. Stilla. 2020. "Improving Progress Monitoring by Fusing Point Clouds, Semantic Data and Computer Vision." *Automation in Construction* 116 (February): 103210. https://doi.org/10.1016/j.autcon.2020.103210.

Bruno, N., F. Rechichi, C. Achille, A. Zerbi, R. Roncella, and F. Fassi. 2020. "Integration of Historical GIS Data in a HBIM System." *International Archives of the Photogrammetry, Remote Sensing and Spatial Information Sciences – ISPRS Archives* 43 (B4): 427–34. https://doi.org/10.5194/isprs-archives-XLIII-B4-2020-427-2020.

Chenaux, A., M. Murphy, G. Keenaghan, J. Jenkins, E. McGovern, and S. Pavia. 2011. "Combining a Virtual Learning Tool and Onsite Study Visits of Four Conservation Sites in Europe." *Geoinformatics FCE CTU* 6: 157–69. https://doi.org/10.14311/gi.6.21.

Croce, V., G. Caroti, L. De Luca, K. Jacquot, A. Piemonte, and P. Véron. 2021. "From the Semantic Point Cloud to Heritage-Building Information Modeling: A Semiautomatic Approach Exploiting Machine Learning." *Remote Sensing* 13 (3): 1–34. https://doi.org/10.3390/rs13030461.

Curry, Edward, James O'Donnell, Edward Corry, Souleiman Hasan, Marcus Keane, and Seán O'Riain. 2013. "Linking Building Data in the Cloud: Integrating Cross-Domain Building Data Using Linked Data." *Advanced Engineering Informatics* 27 (2): 206–19. https://doi.org/10.1016/j.aei.2012.10.003.

Dore, C., and M. Murphy. 2012. "Integration of Historic Building Information Modeling (HBIM) and 3D GIS for Recording and Managing Cultural Heritage Sites." In *2012 18th International Conference on Virtual Systems and Multimedia*, 369–76. IEEE. https://doi.org/10.1109/VSMM.2012.6365947.

Dore, Conor, and Maurice Murphy. 2013a. "Semi-Automatic Modelling of Building Façades with Shape Grammars Using Historic Building Information Modelling." *Proceedings of the International Archives of the Photogrammetry, Remote Sensing and Spatial Information Sciences – 3D Virtual Reconstruction and Visualization of Complex Architectures* XL (February): 57–64. https://doi.org/10.5194/isprsarchives-XL-5-W1-57-2013.

Dore, C., and M. Murphy. 2013b. "Semi-Automatic Techniques for as-Built BIM Façade Modeling of Historic Buildings." *In 2013 Digital Heritage International Congress (DigitalHeritage)* 1: 473–80. IEEE. https://doi.org/10.1109/DigitalHeritage.2013.6743786.

Dore, C., and M. Murphy. 2014. "Semi-Automatic Generation of As-Built BIM Façade Geometry from Laser and Image Data." *Journal of Information Technology in Construction ITCON, ITcon19*, 20–46, http://www.itcon.org/2014/2, 20–46.

Dore, C., and M. Murphy. 2017. "Current State of the Art Historic Building Information Modelling." *International Archives of the Photogrammetry, Remote Sensing and Spatial Information Sciences – ISPRS Archives* 42 (2W5): 185–92. https://doi.org/10.5194/isprs-archives-XLII-2-W5-185-2017.

Dore, C., M. Murphy, S. McCarthy, F. Brechin, C. Casidy, and E. Dirix. 2015. "Structural Simulations and Conservation Analysis – Historic Building Information Model (HBIM)." *ISPRS – International Archives of the Photogrammetry, Remote Sensing and Spatial Information Sciences* XL-5/W4 (5W4): 351–57. https://doi.org/10.5194/isprsarchives-XL-5-W4-351-2015.

Empler, Tommaso. 2018. "Information Modeling Procedure to Represent a Territory Affected by Earthquake." *October* 1997: 147–56. https://doi.org/10.26375/disegno.2.2018.16.

Empler, Tommaso, Michele Calvano, and Adriana Caldarone. 2018. "ARIM Procedure for Risk Prevention in Historic Buildings." *40° Convegno Internazionale Dei Docenti Delle Discipline Della Rappresentazione*, no. September. https://www.researchgate.net/publication/327824878_ARIM_procedure_for_risk _prevention_in_historic_buildings.

Empler, T., Calvano, M. and Caldarone, A. 2019. "ARIM for the prevention of seismic risk." *Disegnare Idee Immagini-Ideas Images*, 30(59): 70–81.

Fai, S., and M. Sydor. 2013. "Building Information Modelling and the Documentation of Architectural Heritage: Between the 'typical' and the 'Specific'." *In 2013 Digital Heritage International Congress (DigitalHeritage)* 1: 731–34. IEEE. https://doi.org/10.1109/DigitalHeritage.2013.6743828.

Garagnani, S., and A.M. Manferdini. 2013. "Parametric Accuracy: Building Information Modeling Process Applied To the Cultural Heritage Preservation." *The International Archives of the Photogrammetry, Remote Sensing and Spatial Information Sciences* XL-5/W1 (February): 87–92. https://doi.org/10.5194/isprsarchives-xl-5-w1-87-2013.

Historic England. 2008. "Conservation Principles, Policies and Guidance." https://content.historicengland.org.uk/images-books/publications/conservation-principles-sustainable-management-historic-environment/conservationprinciplespoliciesguidanceapr08web.pdf/.

Hu, Shushan, Edward Corry, Edward Curry, William J.N. Turner, and James O'Donnell. 2016. "Building Performance Optimisation: A Hybrid Architecture for the Integration of Contextual Information and Time-Series Data." *Automation in Construction* 70 (October): 51–61. https://doi.org/10.1016/j.autcon.2016.05.018.

Idjaton, K., X. Desquesnes, S. Treuillet, and X. Brunetaud. 2021. "Stone-by-Stone Segmentation for Monitoring Large Historical Monuments Using Deep Neural Networks." *Lecture Notes in Computer Science (Including Subseries Lecture Notes in Artificial Intelligence and Lecture Notes in Bioinformatics), 12667 LNCS*, 235–248. https://doi.org/10.1007/978-3-030-68787-8_17.

Isailović, D., V. Stojanovic, M. Trapp, R. Richter, R. Hajdin, and J. Döllner. 2020. "Bridge Damage: Detection, IFC-based Semantic Enrichment and Visualization." *Automation in Construction* 112 (May 2019): 103088. https://doi.org/10.1016/j.autcon.2020.103088.

Khalil, A., and S. Stravoravdis. 2019. "H-BIM and the Domains of Data Investigations of Heritage Buildings Current State of the Art." *ISPRS Annals of the Photogrammetry, Remote Sensing and Spatial Information Sciences* 42 (2/W11): 661–667. https://doi.org/10.5194/isprs-Archives-XLII-2-W11-661-2019.

Khodeir, L.M., D. Aly, and S. Tarek. 2016. "Integrating HBIM (Heritage Building Information Modeling) Tools in the Application of Sustainable Retrofitting of Heritage Buildings in Egypt." *Procedia Environmental Sciences* 34: 258–270. https://doi.org/10.1016/j.proenv.2016.04.024.

Kim, Seungho, Sangyong Kim, and Dong-Eun Lee. 2020. "Sustainable Application of Hybrid Point Cloud and BIM Method for Tracking Construction Progress." *Sustainability* 12 (10): 4106. https://doi.org/10.3390/su12104106.

Koo, B., R. Jung, and Y. Yu. 2021. "Automatic Classification of Wall and Door BIM Element Subtypes Using 3D Geometric Deep Neural Networks." *Advanced Engineering Informatics* 47 (November 2020): 101200. https://doi.org/10.1016/j.aei.2020.101200.

La Russa, F.M., and C. Santagati. 2020. "Historical Sentient – Building Information Model: A Digital Twin for the Management of Museum Collections in Historical Architectures." *International Archives of the Photogrammetry, Remote Sensing and Spatial Information Sciences – ISPRS Archives* 43(B4): 755–762. https://doi.org/10.5194/isprs-archives-XLIII-B4-2020-755-2020.

Lee, J., J. Kim, J. Ahn, and W. Woo. 2019. "Context-Aware Risk Management for Architectural Heritage Using Historic Building Information Modeling and Virtual Reality." *Journal of Cultural Heritage* 38: 242–252. https://doi.org/10.1016/j.culher.2018.12.010.

León-Robles, C.A., J.F. Reinoso-Gordo, and J.J. González-Quiñones. 2019. "Heritage Building Information Modeling (H-BIM) Applied to a Stone Bridge." *ISPRS International Journal of Geo-Information* 8 (3). https://doi.org/10.3390/ijgi8030121

Logothetis, S., A. Delinasiou, and E. Stylianidis. 2015. "Building Information Modelling for Cultural Heritage: A Review." *ISPRS Annals of the Photogrammetry, Remote Sensing and Spatial Information Sciences* 2 (5W3): 177–183. https://doi.org/10.5194/isprsannals-II-5-W3-177-2015.

Lopez-Gonzalez, Laura, Raquel Otero de Cosca, Miguel Gomez-Heras, and Soledad Garcia-Morales. 2016. "A 4D GIS Methodology to Study Variations in Evaporation Points on a Heritage Building." *Environmental Earth Sciences* 75 (14): 1–9. https://doi.org/10.1007/s12665-016-5907-8.

López, F.J., P.M. Lerones, J. Llamas, J. Gómez-García-Bermejo, and E. Zalama. 2017. "A Framework for Using Point Cloud Data of Heritage Buildings Toward Geometry Modeling in A BIM Context: A Case Study on Santa Maria La Real De Mave Church." *International Journal of Architectural Heritage* 11 (7): 965–986. https://doi.org/10.1080/15583058.2017.1325541.

Ma, J.W., T. Czerniawski, and F. Leite. 2020. "Semantic Segmentation of Point Clouds of Building Interiors with Deep Learning: Augmenting Training Datasets with Synthetic BIM-Based Point

Clouds." *Automation in Construction* 113 (September 2019): 103144. https://doi.org/10.1016/j.autcon.2020.103144.

Macwilliam, K., and C. Nunes. 2019. "Structural Analysis of Historical Constructions." 18 (January): 1949–1958. https://doi.org/10.1007/978-3-319-99441-3.

Maietti, F., R. Di Giulio, E. Piaia, M. Medici, and F. Ferrari. 2018. "Enhancing Heritage Fruition Through 3D Semantic Modelling and Digital Tools: The Inception Project." *IOP Conference Series: Materials Science and Engineering* 364 (1). https://doi.org/10.1088/1757-899X/364/1/012089.

Mania, K., A. Psalti, D.M. Lala, M. Tsakoumaki, A. Polychronakis, A. Rempoulaki, ... E. Maravelakis. 2021. "Combining 3D Surveying with Archaeological Uncertainty: The Metopes of the Athenian Treasury at Delphi." In *2021 12th International Conference on Information, Intelligence, Systems & Applications (IISA)*, 1–4. IEEE, July.

Maravelakis, E., A. Konstantaras, J. Kilty, E. Karapidakis, and E. Katsifarakis. 2014a. "Automatic Building Identification and Features Extraction from Aerial Images: Application on the Historic 1866 Square of Chania Greece." *2014 International Symposium on Fundamentals of Electrical Engineering (ISFEE)*, 1–6. https://doi.org/10.1109/ISFEE.2014.7050594.

Maravelakis, E., A. Konstantaras, K. Kabassi, I. Chrysakis, C. Georgis, and A. Axaridou. 2014b. "3DSYSTEK Web-Based Point Cloud Viewer." In *IISA 2014, The 5th International Conference on Information, Intelligence, Systems and Applications*, 260–266. https://doi.org/10.1109/IISA.2014.6878726.

Martínez-Carricondo, P., F. Carvajal-Ramírez, L. Yero-Paneque, and F. Agüera-Vega. 2020. "Combination of Nadiral and Oblique UAV Photogrammetry and HBIM for the Virtual Reconstruction of Cultural Heritage. Case Study of Cortijo del Fraile in Níjar, Almería (Spain)." *Building Research and Information* 48 (2): 140–159. https://doi.org/10.1080/09613218.2019.1626213.

Marzouk, Mohamed. 2020. "Using 3D Laser Scanning to Analyze Heritage Structures: The Case Study of Egyptian Palace." *Journal of Civil Engineering and Management* 26 (1): 53–65. https://doi.org/10.3846/jcem.2020.11520.

Marzouk, Mohamed, and Ahmed Abdelaty. 2014. "Monitoring Thermal Comfort in Subways Using Building Information Modeling." *Energy and Buildings* 84 (December): 252–57. https://doi.org/10.1016/j.enbuild.2014.08.006.

Mazairac, Wiet, and Jakob Beetz. 2013. "BIMQL – An Open Query Language for Building Information Models." *Advanced Engineering Informatics* 27 (4): 444–56. https://doi.org/10.1016/j.aei.2013.06.001.

McArthur, J.J., Nima Shahbazi, Ricky Fok, Christopher Raghubar, Brandon Bortoluzzi, and Aijun An. 2018. "Machine Learning and BIM Visualization for Maintenance Issue Classification and Enhanced Data Collection." *Advanced Engineering Informatics* 38 (June): 101–12. https://doi.org/10.1016/j.aei.2018.06.007.

Morbidoni, C., R. Pierdicca, M. Paolanti, R. Quattrini, and R. Mammoli. 2020. "Learning from Synthetic Point Cloud Data for Historical Buildings Semantic Segmentation." *Journal on Computing and Cultural Heritage* 13 (4). https://doi.org/10.1145/3409262.

Motsa, S.M., G.A. Drosopoulos, M.E. Stavroulaki, E. Maravelakis, R.P. Borg, P. Galea, S. d'Amico, … G.E. Stavroulakis. 2020. "Structural Investigation of Mnajdra Megalithic Monument in Malta." *Journal of Cultural Heritage* 41: 96–105. Cited 3 times. http://www.elsevier.com. doi:10.1016/j.culher.2019.07.004.

Murphy, Maurice. 2012. *Historic Building Information Modelling (HBIM) PhD.* Dublin Institute of Technology. https://doi.org/10.1108/02630800910985108.

Murphy, Maurice, Eugene McGovern, and Sara Pavia. 2009. "Historic Building Information Modelling (HBIM)." *Structural Survey* 27 (4): 311–27. https://doi.org/10.1108/02630800910985108.

Murphy, M., E. Meegan, G. Keenaghan, A. Chenaux, A. Corns, S. Fai, L. Chow, et al. 2021. "Shape Grammar Libraries of European Classical Architectural Elements for Historic BIM." *The International Archives of the Photogrammetry, Remote Sensing and Spatial Information Sciences* XLVI-M-1-2 (September): 479–86. https://doi.org/10.5194/isprs-archives-XLVI-M-1-2021-479-2021.

Oreni, D., R. Brumana, A. Georgopoulos, and B. Cuca. 2013. "HBIM for Conservation and Management of Built Heritage: Towards a Library of Vaults and Wooden Bean Floors." *ISPRS Annals of*

Photogrammetry, Remote Sensing and Spatial Information Sciences II-5/W1 (September): 215–21. https://doi.org/10.5194/isprsannals-II-5-W1-215-2013.

Pierdicca, R., M. Paolanti, F. Matrone, M. Martini, C. Morbidoni, E.S. Malinverni, E. Frontoni, and A.M. Lingua. 2020. "Point Cloud Semantic Segmentation Using a Deep Learning Framework for Cultural Heritage." *Remote Sensing* 12 (6): 1–23. https://doi.org/10.3390/rs12061005.

Pocobelli, D.P., J. Boehm, P. Bryan, J. Still, and J. Grau-Bové. 2018. "BIM for Heritage Science: A Review." *Heritage Science* 6 (1). https://doi.org/10.1186/s40494-018-0191-4.

Pocobelli, D.Ph., J. Grau-Bové, J. Boehm, P. Bryan, and J. Still. 2021. *Heritage Building Information Model (BIM) for Scientific Data.* Doctoral thesis (Ph.D), UCL (University College London).

Pour Rahimian, Farzad, Saleh Seyedzadeh, Stephen Oliver, Sergio Rodriguez, and Nashwan Dawood. 2020. "On-Demand Monitoring of Construction Projects through a Game-like Hybrid Application of BIM and Machine Learning." *Automation in Construction* 110 (October 2019): 103012. https://doi.org/10.1016/j.autcon.2019.103012.

Rinaudo, F., E. Agosto, and P. Ardissone. 2007. "GIS and Web-GIS, Commercial and Open Source Platforms: General Rules for Cultural Heritage Documentation." *International Archives of the Photogrammetry, Remote Sensing and Spatial Information Sciences* XXXVI-5/C5: 625–30.

Romero-Jarén, R., and J.J. Arranz. 2021. "Automatic Segmentation and Classification of BIM Elements from Point Clouds." *Automation in Construction* 124 (December 2020). https://doi.org/10.1016/j.autcon.2021.103576.

Saygi, G., G. Agugiaro, M. Hamamcıoğlu-Turan, and F. Remondino. 2013. "Evaluation of GIS and BIM Roles for the Information Management of Historical Buildings." *ISPRS Annals of the Photogrammetry, Remote Sensing and Spatial Information Sciences* II (September): 283–88. https://doi.org/10.5194/isprsannals-II-5W1-283-2013.

Seghier, Taki Eddine, Yaik Wah Lim, Mohd Hamdan Ahmad, and Williams Opeyemi Samuel. 2017. "Building Envelope Thermal Performance Assessment Using Visual Programming and BIM, Based on ETTV Requirement of Green Mark and GreenRE." *International Journal of Built Environment and Sustainability* 4 (3): 227–35. https://doi.org/10.11113/ijbes.v4.n3.216.

Sifniotis, Maria, P. Watten, K. Mania, and M. White. 2007. "Influencing Factors on the Visualisation of Archaeological Uncertainty." In *VAST: International Symposium on Virtual Reality, Archaeology and Intelligent Cultural Heritage*, 79–85.

Sifniotis, M., B. Jackson, K. Mania, N. Vlassis, P.L. Watten, and M. White. 2010. "3D Visualization of Archaeological Uncertainty." In *Proceedings – APGV 2010: Symposium on Applied Perception in Graphics and Visualization*, 162. https://doi.org/10.1145/1836248.1836284.

Simeone, Davide. 2018. "BIM and Behavioural Simulation for Existing Buildings Re-Use Design." *Tema: Technology, Engineering, Materials and Architecture* 4 (2): 59–69.

Sketchfab. n.d. "Sketchfab." Accessed June 10, 2022. https://sketchfab.com.

Soilán, M., A. Justo, A. Sánchez-Rodríguez, and B. Riveiro. 2020. "3D Point Cloud to BIM: Semi-Automated Framework to Define IFC Alignment Entities from MLS-Acquired LiDAR Data of Highway Roads." *Remote Sensing* 12 (14). https://doi.org/10.3390/rs12142301.

Stiny, George, and James Gips. 1972. "Shape Grammars and the Generative Specification of Painting and Sculpture." In *Information Processing 71 Proceedings of the IFIP Congress 1971.* Volume 2. https://doi.org/citeulike-article-id:1526281.

Tang, Shu, Dennis R. Shelden, Charles M. Eastman, Pardis Pishdad-Bozorgi, and Xinghua Gao. 2019. "A Review of Building Information Modeling (BIM) and the Internet of Things (IoT) Devices Integration: Present Status and Future Trends." *Automation in Construction* 101 (January): 127–39. https://doi.org/10.1016/j.autcon.2019.01.020.

Tapkın, S., E. Tercan, S.M. Motsa, G. Drosopoulos, M. Stavroulaki, E. Maravelakis, and G. Stavroulakis. 2022. "Structural Investigation of Masonry Arch Bridges Using Various Nonlinear Finite-Element Models." *Journal of Bridge Engineering* 27 (7): 04022053.

Tkáč, M., P. Mesároš, and T. Mandičák. 2018. "Terrestrial Laser Scanning - Effective Technology for Creating Building Information Models." *Pollack Periodica* 13 (3): 61–72. https://doi.org/10.1556/606.2018.13.3.7.

"Tridify" (2022). www.tridify.com. Last accessed: 2022-06-10

Valero, E., A. Forster, F. Bosché, C. Renier, E. Hyslop, and L. Wilson. 2018. "High Level-of-Detail BIM and Machine Learning for Automated Masonry Wall Defect Surveying." In *ISARC 2018 – 35th International Symposium on Automation and Robotics in Construction and International AEC/FM Hackathon: The Future of Building Things*. https://doi.org/10.22260/isarc2018/0101.

Worrell, Laura Lee, "Building Information Modeling (BIM): The Untapped Potential for Preservation Documentation and Management" (2015). All Theses. 2146. https://tigerprints.clemson.edu/all_theses/2146.

Chapter 3

Modeling and Analysis of Injection Molding Process: A Case Study

Thomas Kestis,[1,*] Anastasios Tzotzis,[2] Dimitrios Tzetzis,[1] Panagiotis Kyratsis[2]

[1] Digital Manufacturing and Materials Characterization Laboratory, School of Science and Technology, International Hellenic University, N. Moudania, Greece;

[2] Department of Product and Systems Design Engineering, University of Western Macedonia, Kila, Kozani, Greece

Abstract

Nowadays, the simulation of manufacturing processes with the use of specialized software is a common method for determining potential defects and optimizing the process parameters. Regarding the plastic industries, the simulation of the injection molding process is a crucial step before moving to the production stage.

In this study, a plastic part is examined through the simulation of the injection molding process. The part has a design flaw that results in a manufacturing defect. In particular, an instance of the plastic shell of a 2 × 2 rotational cubic puzzle develops a distinctive sink mark. SolidWorks™ Plastics, a specialized CAD/CAE simulation software used for injection molding, is utilized in order to simulate the process with the actual operational conditions and verify the defect. The design flaws and/or other factors that are responsible for the development of the defect are identified.

The process parameters of the simulation are examined. The most influential parameters for the development of the sink mark are selected for further analysis through a number of injection molding simulations. The optimization of the parameters and its potential effect on the reduction of the sink mark is discussed.

Keywords: injection molding, process parameters, defects, sink marks, SolidWorks™ Plastics

Introduction

Injection molding is a manufacturing process for producing parts by injecting melt material into the cavity of a mold. Many products such as mechanical and automotive parts, toys, storage containers, and other plastic products mostly available today are created by injection molding. Injection molding is the most common process of producing plastic products.

---

* Corresponding Author Email: thomas2002gr@yahoo.gr

Similar to other manufacturing processes, injection molding routinely produces defective parts. They can be caused either by defects in the molds or by the molding process itself. In situations where the molding parameters are not controlled properly, a variety of common defects may be caused, such as dimensional variations, sink marks, voids, weld/meld lines, poor surface finish, air traps, and burn marks. Of all attribute defects, sink marks are considered to be the most difficult to deal with.

Injection molding simulation with the use of software tools is a common method for determining potential defects and optimizing the process parameters. Hence, a great number of plastics industries can simulate the injection molding process before getting to the production stage, maximize the efficiency of the production cycle, and minimize overall manufacturing costs.

A number of studies are available in the literature. Yu et al. [1] studied injection molding of thin plates, on a micro-scale, in order to manufacture micro-fluidic devices for bioMEMS applications. The authors found that the injection speed and mold temperature in injection molding greatly affect the replication accuracy. Additionally, with the aid of 2D simulations the data were compared with the equivalent experimental ones. Zema et al. [2] presented the issues and advantages connected with the transfer of injection molding from the plastics industry to the production of conventional and controlled-release dosage forms. German [3] addressed the four critical factors that are required to successfully use titanium powder in injection molding, with an emphasis on demanding applications in aerospace and medical fields. In the work of Fernandes et al. [4], a review of the research done in the field of modeling and optimization during injection molding was presented. Specifically, the modeling and optimization techniques were discussed, stating both the advantages and disadvantages of each one of the most used methods. Tzetzis et al. [5–6] simulated the process of performing runner balancing for the two uneven cavities of a product. The authors additionally examined the effects of standard injection molding process parameters on the process. Similarly, Dang [7] reviewed the state of the art of process parameter optimization for plastic injection molding through case studies. The implementation of the suggested frameworks was demonstrated, as well as compared. Staněk et al. [8] described in their work, the Moldflow Plastics Xpert system and its usage for optimization of the various parameters during the injection molding process of real parts at the stage of their production. Zhang et al. [9] demonstrated the manufacturing process of a prototype bulk metallic glass injection molding tool that is capable of producing micro polymeric components, with sub-micron surface features. Xhu et al. [10] proposed a combination of techniques such as neural networks and numerical analysis, to optimize the injection molding process. Furthermore, a number of CAD-based works in similar fields show that CAD systems provide flexible and robust analyzing tools for various cases such as computational design [11], design automation [12–15], and study [16–17].

In the present work, the process parameters of injection molding of a real plastic part are studied, in order to examine their effect on typical defects such as the sink mark size.

Product Description

In this study, a plastic component from the 2 × 2 flat rotational cubic puzzle, produced by Verdes Innovations S.A., or the flat V-CUBE™2, is chosen for a comprehensive design review and mold flow analysis. The final part is decorated with printed colors, a full assembly of the product.

V-CUBE™2 is the smallest member of the V-CUBE™ family. It is available in two versions: the classic flat design and the "pillow" shape. Both designs have the same V-CUBE™ internal mechanism. V-CUBE™2 is a superior quality, multicolored, two-layer cube with smooth rotation and excellent durability. V-CUBE™2 has almost 3.7 million possible permutations and weighs 74 g. The cube consists of eight cubic pieces at the visible corners, and a solid cross for supporting their independent rotation on based axes.

The outer plastic components can be extracted with the application of some force. Then, the assembly interface of the internal mechanism and the internal design of the outer plastic component are revealed (Figure 3.1).

It is evident from Figure 3.2 that the outer plastic components have a sink mark developed close to their corner. The sink mark is even more evident when the light is reflected from the surface of the painted cubes.

Plastic Component Analysis

The bounding box of the plastic component is a cube with 25 mm side. Figure 3.3 depicts the inner configuration of the plastic component.

The visual inspection of the component reveals that the outer shell of the component is relatively thin, but this does not pose a problem to the injection molding process since the overall size of the part is very small.

The inner configuration comprises a set of ribs, a hook, and a boss feature at the base of the hook. The ribs ensure that the part is aligned with the rest of the assembly. The hook ensures that the part stays secured. Of course, the part can be detached from the assembly, but a young child may not be able to achieve it. The purpose of the boss at the base of the hook is to keep the proper offset with the assembly interface of the main mechanism, in order to ensure that the part rotates smoothly and does not collide with the other instances of the outer surface. Nonetheless, it is evident that this feature is the culprit for the development of the defect since the distinctive sink mark develops right at the adjacent side of the part. Figure 3.4 depicts this sink mark.

Figure 3.1 Internal mechanism of the rotational cubic puzzle

Figure 3.2 Sink marks on the surface of the rotational cubic puzzles

Figure 3.3 Inner configuration of the plastic component

CAD Model

The CAD model of the plastic component was provided by the manufacturer in Parasolid file format (.x_t file extension). This is the CAD model used for the actual mold creation. The file was imported into SolidWorks for the purpose of the study. A clipping plane at the level of the boss feature reveals the cause of the sink mark at the outer surface of the part Figure 3.5.

Indeed, the width of the boss feature is 154% bigger than the wall thickness and the depth 66%. Therefore, the geometry of the boss feature consists of a design flaw that is highly probable to end up in the formation of a sink mark.

Injection Molding Process Parameters

The manufacturing of the part is accomplished by using an ARBURG 305-210-700 injection molding machine, capable of delivering up to 150 MPa of pressure, as shown in Figure 3.6.

Figure 3.4 Sink mark on the outer surface of the plastic part

Figure 3.5 Clipping plane of plastic part

The material selected for the fabrication of the part is CHIMEI Polylac PA-757 polystyrene. Polylac PA-757 is an Acrylonitrile Butadiene Styrene (ABS) product for general use. The actual parameters of the injection molding process, as provided by the manufacturer, are listed in Table 3.1, whereas the material properties are displayed in Table 3.2.

Process Simulation

To perform an injection molding simulation and obtain results that can be compared with the actual phenomena, the choice of a proper software solution is critical. In the present study, the software tool of choice is SolidWorks™ Plastics.

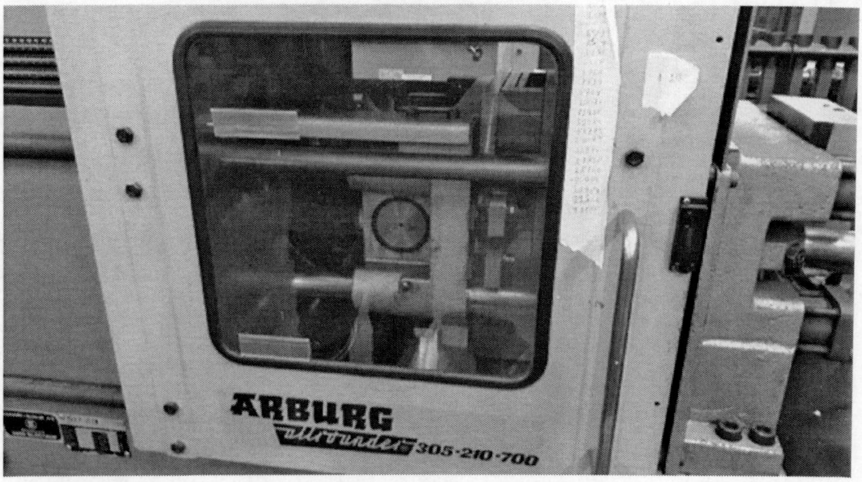

Figure 3.6 Injection molding machine used in the process

Table 3.1 Actual injection molding process parameters

Injection pressure	Up to 150 MPa
Melt temperature	230°C
Mold temperature	70°C
Material	CHI MEI Polylac PA-757
Total cycle time	12.5 sec

Table 3.2 Polylac PA-757 material properties

Max melt temperature	240°C
Min melt temperature	170°C
Max mold temperature	60°C
Min mold temperature	10°C
Ejection temperature	100°C
Glass transition temperature	105°C
Specific heat	2,430 J/(kg K)
Thermal conductivity	0.197 W/(mK)
Elastic modulus	2.84×10^{10} Pa
Poisson's ratio	0.35
Thermal expansion coefficient	9.4×10^{-5}
Max shear rate	49,000 1/s
Max shear stress	297,000 Pa

Manufacturers of products with injection-molded parts can resolve design and tooling challenges by performing accurate mold-filling simulations using SolidWorks Plastics software. Rather than rely on time-consuming and costly prototyping and tooling iterations to improve manufacturability, injection molding professionals can utilize this solution to cut time and cost from the process while simultaneously improving quality.

Mesh Generation

The first stage of the simulation preparation consists of the part's meshing. The meshing procedure is of critical importance for the accuracy of the simulation.

In order to perform the simulation, the CAD geometry needs to be converted to a network of interconnecting triangular elements that accurately represent the actual geometry of the part that will not participate in the simulation anymore. All the calculations occur thereafter on the nodes of the interconnecting triangular elements. In this process, the part is identified as cavity, insert, runner, or mold. This analysis is a cavity analysis. That means that the melted plastic material will flow through the volume of all the parts identified as cavities during the injection molding process.

The mesh is refined by reducing the triangle size and applying local refinement. The refinement is performed automatically in regions with small features or high curvatures. The mesh tolerance defines the minimum allowable element size for automatic refinement and is given by the value of triangle size multiplied by the value of mesh tolerance. Accepted values are between the range of 0.1 and 1.0. In this study, the default value equal to 0.3 is used. Finally, a hybrid mesh type is selected which leads to the result seen in Figure 3.7. The cutaway view shows the transition between the elements of the surface and the core of the part.

Process Parameters

SolidWorks™ Plastics estimates the part's volume and provides the user with a recommendation on how long it will take the molten plastic to be injected and completely fill the mold. Melt and mold temperature boxes actually contain the resin manufacturer's recommendations on how the user should operate the injection molding machine with this particular material. For this study, the "Filling time" is accepted with the default calculated value (0.9 s) and the other parameters are adjusted according to the values given by the manufacturer.

The goal of packing is to produce a part with uniform weight and dimensional integrity. Successful packing improves the part quality. During the first stage of packing, pressure is applied to the injection system as the molten plastic in the mold cools and shrinks. The pressure forces additional material into the mold to compensate for thermal shrinkage. During the

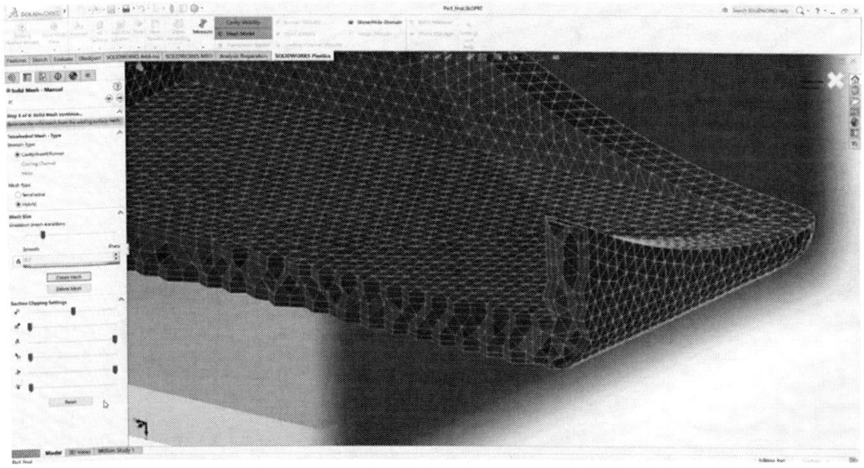

Figure 3.7 Hybrid mesh element transition

last stage of packing, the "pure cooling" stage, the injection pressure is removed, and only the temperature of the part is calculated as the freezing completes.

The software estimates "pressure holding time" (2.77 s) required for the packing stage and "cooling time" (6.24 s) required for the pure cooling stage during the simulation process.

Polymer material at the specified melt temperature is introduced into the cavity through injection locations. The injection location of the part can be identified from a visual inspection. Therefore, it is introduced at the node with the closest proximity to the injection location of the actual part. The pointer diameter is set to 1 mm, in order to match the actual pointer diameter as given by the manufacturer (Figure 3.8).

Results and Discussion

In the first part of the study, SolidWorks™ Plastics was utilized to simulate the injection molding process of an actual injection-molded plastic part and compare the mold-filling simulation predictions with the actual results.

The injection molding process of the actual part was confirmed, as the simulation results are in very good agreement with the actual process. The cavity of the part is easily filled with an injection pressure that is significantly lower than the maximum limit specified for the injection molding machine. The maximum temperature at end of fill has remained within 10°C of the starting melt temperature and predicts little to no risk of plastic material degradation. The simulation of our study predicts no other defects, such as short shots or air traps, than the distinctive sink mark at the outer surface of the part. The location and magnitude of the predicted sink mark are virtually identical to the actual part. Figure 3.9 illustrates the comparison between the predicted sink mark and the surface defect of the actual part.

The second part of the study includes a number of simulation runs that are used to analyze the injection molding parameters, which lead to the elimination of the defect of the part.

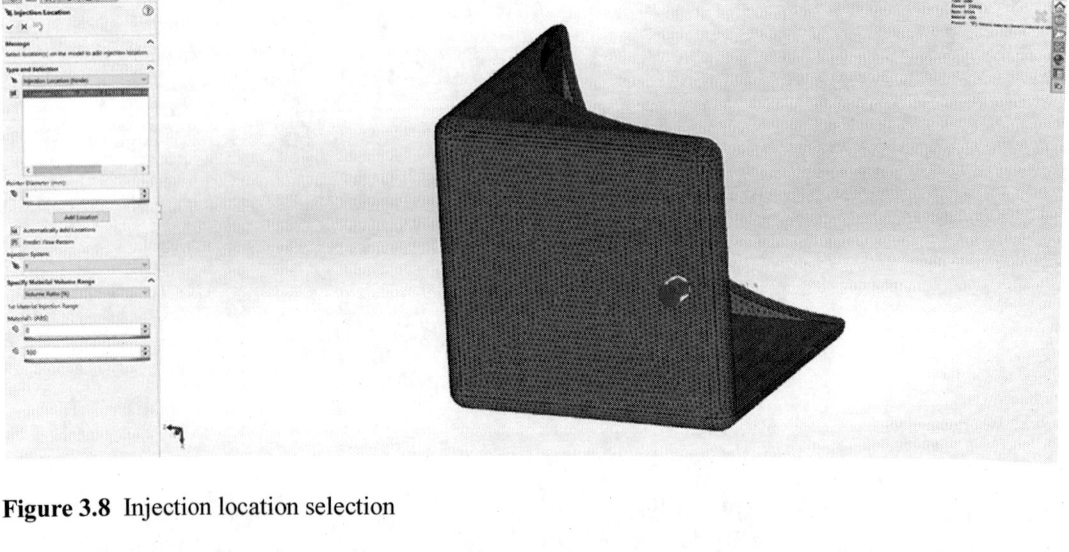

Figure 3.8 Injection location selection

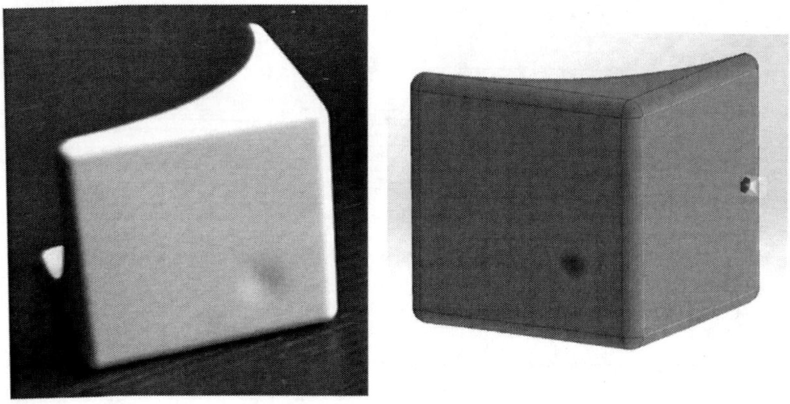

Figure 3.9 Comparison between the actual part and the simulated one

Table 3.3 Factors and their levels used in the study

Level	T_{melt} (°C)	T_{mold} (°C)	$t_{pressure\ holding}$ (s)
−1	210	60	2.27
0	230	70	2.77
+1	245	80	3.27

Design of Experiments

Similar works in the field [18–19] reported that the following factors are observed to affect the development of sink marks the most and hence are selected for further study:

- Melt temperature
- Mold temperature
- Pressure holding time

Other factors such as part geometry, material, packing pressure, and gate location are considered constants. For each of the factors, three levels (−1, 0, +1) were used. The level settings for the mold temperature and the melt temperature are selected after discussion with the mold makers of the part manufacturer. The estimated value from the evaluation simulation for the pressure holding time is selected as the medium (0) level for reference. Two other values, a lower (−1) and a higher one (+1), were then chosen in order to examine the effect of this factor. The selected values of the process parameters are summarized in Table 3.3.

The impact of the parameters was examined with the conduction of a number of simulations. For each parameter, three simulations were carried out, one for each level. The other parameters were kept constant and retained the values given by the manufacturer, as attributed in the evaluation simulation of the first part of this study, leading to a total of seven runs (Table 3.4).

Melt Temperature Effect

In the first simulation, the melt temperature is set to the low value of 210°C. It represents the lower recommended value that may be used in the actual injection molding process. The value

Table 3.4 Simulation parameters per run

Run	T_{melt} (°C)	T_{mold} (°C)	$t_{pressure\ holding}$ (s)	$D_{sink\ mark}$ (mm)
1	210	70	2.77	0.0817
2	230	70	2.77	0.0886
3	245	70	2.77	0.0881
4	230	60	2.77	0.0903
2	230	70	2.77	0.0886
5	230	80	2.77	0.0821
6	230	70	2.27	0.0886
2	230	70	2.77	0.0886
7	230	70	3.27	0.0886

was selected after a discussion with the part manufacturer. It is of interest to note that according to the polymer material parameters chart, the lower melt temperature may be set at 170°C. Nonetheless, the lower recommended temperature for the actual process is considerably higher. The reason may be that 170°C is very low, so the melt solidifies quickly, and this increases the risk of developing unacceptable part defects, such as short shots. The value of the sink mark at low melt temperature is 0.081712 mm. The value of the maximum predicted sink mark at the simulation of the actual injection molding process of the part was 0.0886 mm. The selection of the lower recommended value for the melt temperature resulted in a reduction of 8% of the predicted sink mark. Lower melt temperatures result in higher viscosities, which make it difficult for the melt to flow while filling the mold cavity. Nonetheless, no other defects developed in the simulation, especially short shots.

In the second simulation, the melt temperature is set to the medium value of 230°C. This is the recommended value for the actual injection molding process of the part as provided by the manufacturer. The value of the predicted sink mark at medium melt temperature is 0.0886 mm.

In the third simulation, the melt temperature is set to the high value of 245°C. It represents the higher value that may be used in the actual injection molding process as given by the part manufacturer. According to the polymer material parameters chart, the higher melt temperature may be set at 240°C. Nonetheless, the higher melt temperature as given by the part manufacturer is 5°C above the recommended maximum by the polymer material chart. It is not uncommon that plastic manufacturers rely on experience and, depending on the occasion, may experiment and use parameters with values that differ or exceed the recommended ranges in order to achieve optimal results.

The value of the sink mark at a high melt temperature is 0.088122 mm. The value of the maximum predicted sink mark at the simulation of the actual injection molding process of the part was 0.0886 mm. The selection of the higher recommended value for the melt temperature resulted in practically the same magnitude of the sink mark. The slight reduction of approximately 0.5% is not of statistical importance. As expected, no short shots developed in the simulation. Higher melt temperatures result in lower viscosities, which make it easier for the melt to fill the mold cavity. However, if the ease of fill of the part does not pose an issue, higher melt temperatures are not recommended because the part needs longer cooling time before it is ejected, and this results in longer cycle time and, thus, reduced overall productivity.

It is observed (Figure 3.10) that the melt temperature is an influential process parameter in the development of the sink mark. This factor needs to be considered as an important input to plastic product manufacturing. The reduction of the sink mark at the low melt temperature is 8% compared to the medium melt temperature, which is used as a reference. This can be explained by the higher viscosity of the melt. The melt needs a shorter cooling time to solidify, and, thus, the resulting sink mark is considerably reduced. At the high melt temperature, the sink mark is practically the same as the reference result. Higher melt temperatures are not recommended unless the ease of fill of the part is an issue. At higher melt temperatures the part needs longer cooling time before it is ejected, and this results in longer cycle time and reduced overall productivity.

Mold Temperature Effect

The value of the sink mark at low mold temperature is 0.090311 mm (Simulation No 4). The value of the maximum predicted sink mark at the simulation of the actual injection molding process of the part was 0.0886 mm. The selection of the lower recommended value for the mold temperature resulted in an increase of 2% of the predicted sink mark.

The behavior of mold temperature is expected to be similar to that of the melt temperature parameter. Lower mold temperatures would result in higher viscosities, the melt would need a shorter cooling time to solidify, and the resulting sink mark would be reduced. However, in this case, the predicted sink mark is increased by 2% compared to the reference simulation with the medium level of mold temperature. It is evident from similar works [18–19] that the effect of mold temperature on the sink mark relies heavily on the geometry of the part, and, depending on the occasion, it may be difficult to predict the behavior of the parameter unless a series of simulations are conducted.

The value of the sink mark at high mold temperature is 0.082146 mm (simulation No 5). The value of the maximum predicted sink mark at the simulation of the actual injection molding process of the part was 0.0886 mm. The selection of the higher recommended value for the mold temperature resulted in a decrease of 7% of the predicted sink mark.

Again, the behavior of the parameter of the mold temperature at the higher temperature is unexpected. Higher mold temperatures would result in lower viscosities, the melt would need longer cooling time to solidify and the resulting sink mark would be increased. However, the sink mark is decreased, and this result demonstrates that the effect of the mold temperature on the sink mark depends on the part geometry. As expected, no short shots developed in the

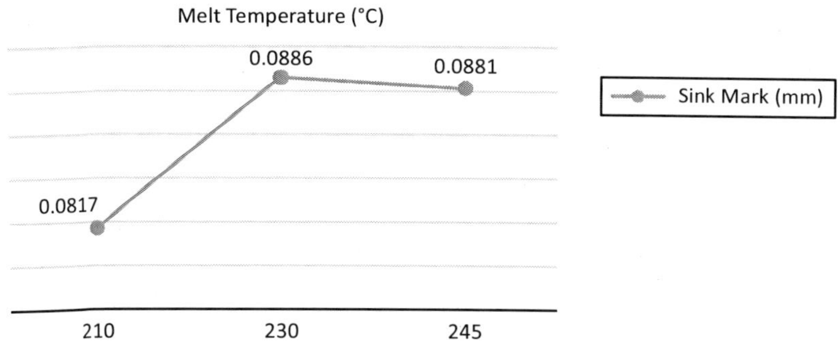

Figure 3.10 Summary of results for melt temperature parameter simulations

simulation. Higher mold temperatures result in lower viscosities, which make it easier for the melt to fill the mold cavity.

From the results analysis (Figure 3.11), it is observed that the mold temperature is a process parameter that affects the development of the sink mark to a level compared to that of the melt temperature. This factor also needs to be considered as an important input to plastic product manufacturing.

At the highest effect, the reduction of the sink mark is 7% compared to the mold temperature used as a reference. Higher mold temperatures would result in lower viscosities, the melt would need longer cooling time to solidify, and the resulting sink mark would be increased. However, the reduction of the sink mark occurred at the high mold temperature, and this is an unexpected result. Also, at the low mold temperature the sink mark, instead of reducing, increased.

Pressure Holding Time Effect

In the sixth simulation, the pressure holding time is set to a low value of 2.27s. The value is set 0.5s lower than the reference level, in order to allow to draw safe conclusions from the results but not deviate significantly from the value used in the actual injection molding process. Also, too short pressure holding time may also lead to reverse flow if the gate does not freeze off.

The value of the predicted sink mark at low pressure holding time is 0.0886 mm. As expected, the decrease in the pressure holding time has not affected the sink mark in the filling stage of the simulation.

The value of the predicted volumetric shrinkage at low pressure holding time is 8.7265%. The value of the maximum predicted volumetric shrinkage at the reference simulation with medium pressure holding was 8.686%. The selection of the lower value for the pressure holding time resulted in an increase of 0.46% of the predicted volumetric shrinkage.

During the packing stage, pressure is applied to the injection system as the molten plastic in the mold cools and shrinks. The pressure forces additional material into the mold to compensate for thermal shrinkage. When the pressure holding time decreases, less material is forced into the mold, so more volumetric shrinkage is expected to occur. Indeed, in our simulation the volumetric shrinkage at the end of the packing stage increased but to a degree that may not be of statistical importance.

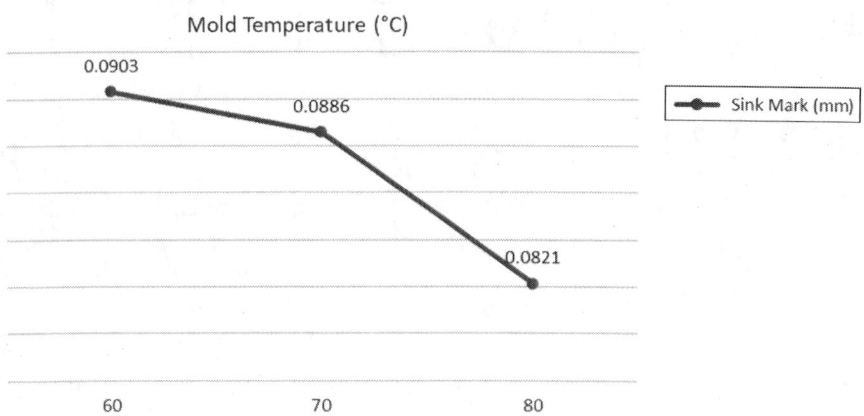

Figure 3.11 Summary of results for mold temperature parameter simulations

In the seventh simulation, the pressure holding time is set to the high value of 3.27s. Since the size of our part is small and the recommended pressure holding time is 2.57s, a total deviation of 1s from the low to the high value should be enough to draw safe conclusions about the effect of the pressure holding time factor. Also, large holding times would need larger cooling times, and this leads to increased total cycle time which may be uneconomical for mass production. As expected, the increase in the pressure holding time has not affected the sink mark in the filling stage of the simulation. The value is 0.0886 mm.

The value of the predicted volumetric shrinkage at high pressure holding time is 8.6133%. The value of the maximum predicted volumetric shrinkage at the reference simulation with medium pressure holding was 8.686% (Figure 3.12). The selection of the higher value for the pressure holding time resulted in a decrease of 0.83% of the predicted volumetric shrinkage.

When the pressure holding time increases, more material is forced into the mold, so less volumetric shrinkage is expected to occur. Indeed, in our simulation the volumetric shrinkage at end of the packing stage decreased but again to a degree that may not be of statistical importance.

The chart depicts the values of volumetric shrinkage at the end of the packing stage for the three values of pressure holding time selected for the analysis. It is expressed as a percentage of shrinkage from the initial volume of the part.

It is observed that deviations in the pressure holding time influence the volumetric shrinkage at the end of the packing stage and, thus, the development of sink marks. The differences, however, of the low and high values in comparison to the reference value are very low.

The reduction of the volumetric shrinkage at the high value of pressure holding time is 0.83%, compared to the reference value, and the increase at the low value of pressure holding time is 0.46%. The results are in good agreement with the expected outcome. That is because the holding pressure at the packing phase compensates to some degree for the material shrinkage of the filling phase. However, the differences are low and indicate that adjustments of the pressure holding time parameter, within the recommended limits, may not be of actual practical value for the reduction of sink marks.

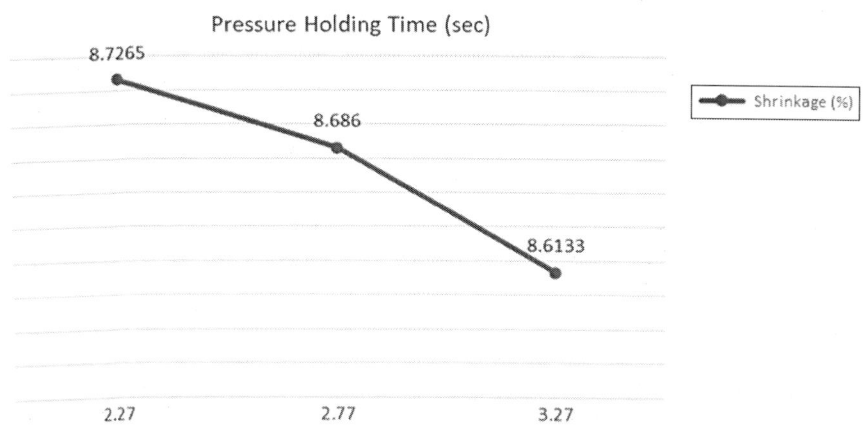

Figure 3.12 Summary of results for pressure holding time parameter simulations

Effect of Combined Factors

From the series of simulations conducted, it is evident that the parameters selected for analysis have an influential effect on the injection molding process and the selection of different values affects the resulting volumetric shrinkage and development of sink marks. When studied in isolation, though, the effect of these factors may be subtle on the final result and in some cases even statistically unimportant. Thus, the practical value of the results may be limited in the actual manufacturing process.

A better approach would be to examine the effect of these factors in combination. In order to conduct the simulation, we pick the values of the factors with the most prominent effect on sink mark and volumetric shrinkage. The selected values of the process parameters for the combined analysis are: T_{melt}=210°C, T_{mold}=80°C, and $t_{pressure\ holding}$=3.27 s. The combined examination of the selected process parameters is expected to affect both the development of the sink mark at the filling stage of the process and the volumetric shrinkage at the packing stage.

The value of the resulting sink mark is 0.077247 mm. The value of the maximum predicted sink mark at the simulation of the actual injection molding process of the part was 0.0886 mm. The combination of melt and mold temperature adjustment resulted in a reduction of 14.6% of the predicted sink mark. It is evident that the adoption of the combined approach results in clearly maximum effect, in regard to the development of the sink mark at the filling stage. It is expected likewise, that the incorporation of the optimal pressure holding time will result in even more reduced volumetric shrinkage at the packing stage of the simulation. The value of the predicted volumetric shrinkage is 7.1116%. The value of the maximum predicted volumetric shrinkage at the reference simulation with medium pressure holding was 8.686%. The combined parameters resulted in a decrease of 1.57% of the predicted volumetric shrinkage. Indeed, the reduction of the volumetric shrinkage is clearly intensified with the adoption of the parameters combined, than with any of the parameters in isolation.

Conclusion

Injection molding is a complex manufacturing process with possible production flaws. The use of software tools is a common method for determining potential defects and optimizing the process parameters. Hence, the value of injection molding simulations in the industry is very important due to minimization of manufacturing costs and delays in production.

In the current study, SolidWorks™ Plastics was used in order to examine the plastic shell of a 2x2 rotational cubic puzzle. A design flaw of the actual part that results in a distinctive sink mark was identified. The CAD model used for mold development and the actual operational conditions were replicated within the software. The simulation analysis verified the defect of the plastic part with accuracy. The results highlighted the importance of simulating the process before proceeding to mold development in order to minimize potential flaws in the actual molding process.

The most influential process parameters on the development of sink marks were selected for further analysis, namely melt temperature, mold temperature, and pressure holding time. A set of simulation experiments was designed in order to examine their effect on the development of the sink mark. Three levels of values were selected for each experiment: low, medium, and high. It was concluded that proper selection of the process parameter values can affect the result and that the adoption of the parameters combined intensified the result, in comparison with the parameters in isolation.

Nonetheless, even with combined process parameters, the development of the sink mark was modestly treated. The result highlights the importance of the proper design of parts for injection molding. If design flaws are not identified in time through simulation or cannot be excluded, there is a high probability that the process will result in defects that may not be possible to eliminate with the optimization of the processing parameters.

References

[1] Yu, L., Koh, C. G., Lee, L. J., Koelling, K. W., & Madou, M. J. (2002). Experimental investigation and numerical simulation of injection molding with micro-features. *Polymer Engineering & Science*, 42(5), 871–888.

[2] Zema, L., Loreti, G., Melocchi, A., Maroni, A., & Gazzaniga, A. (2012). Injection molding and its application to drug delivery. *Journal of Controlled Release*, 159(3), 324–331.

[3] German, R. M. (2013). Progress in titanium metal powder injection molding. *Materials*, 6(8), 3641–3662.

[4] Fernandes, C., Pontes, A. J., Viana, J. C., & Gaspar-Cunha, A. (2018). Modeling and optimization of the injection-molding process: A review. *Advances in Polymer Technology*, 37(2), 429–449.

[5] Tzetzis, D., Sofianidis, I., & Kyratsis, P. (2015). Injection molding simulation for part manufacture with polystyrene. *Applied Mechanics and Materials*, 809–810, 229–234.

[6] Tzetzis, D., Sofianidis, I., & Kyratsis, P. (2015). Computer aided injection multi cavity molding process analysis of thermoplastic polymers. *Applied Mechanics and Materials*, 808, 107–112.

[7] Dang, X. P. (2014). General frameworks for optimization of plastic injection molding process parameters. *Simulation Modelling Practice and Theory*, 41, 15–27.

[8] Staněk, M., Maňas, D., Maňas, M., & Šuba, O. (2011). Optimization of injection molding process. *International Journal of Mathematics and Computers in Simulation*, 5(5), 413–421.

[9] Zhang, N., Byrne, C. J., Browne, D. J., & Gilchrist, M. D. (2012). Towards nano-injection molding. *Materials Today*, 15(5), 216–221.

[10] Xu, Y., Zhang, Q., Zhang, W., & Zhang, P. (2015). Optimization of injection molding process parameters to improve the mechanical performance of polymer product against impact. *The International Journal of Advanced Manufacturing Technology*, 76(9), 2199–2208.

[11] Kyratsis, P., Tzotzis, A., & Manavis, A. (2021). Computational design and digital fabrication. In: Kumar, S., & Rajurkar, K. P. (eds), *Advances in Manufacturing Systems. Lecture Notes in Mechanical Engineering*. Springer, Singapore, 1–16.

[12] Kyratsis, P., Gabis, E., Tzotzis, A., Tzetzis, D., & Kakoulis, K. (2019). CAD based product design: A case study. *International Journal of Modern Manufacturing Technologies*, 11(3), 88–93.

[13] Tzotzis, A., Manavis, A., Efkolidis, N., & Kyratsis, P. (2021). CAD-based automated G-code generation for drilling operations. *International Journal of Modern Manufacturing Technologies*, 13(3), 177–184.

[14] Tzotzis, A., García-Hernández, C., Huertas-Talón, J. L., Tzetzis, D., & Kyratsis, P. (2017). Engineering applications using CAD based application programming interface. *In Proceedings of the MATEC Web of Conferences*, 94, 1–7.

[15] Tzotzis, A., García-Hernández, C., Huertas-Talón, J. L., & Kyratsis, P. (2020). CAD-based automated design of fea-ready cutting tools. *Journal of Manufacturing and Materials Processing*, 4(4), 1–14.

[16] Kyratsis, P., Kakoulis, K., & Markopoulos, A. P. (2020). Advances in CAD/CAM/CAE technologies. *Machines*, 8(1), 13.

[17] Iatrou, G., Tzotzis, A., Kyratsis, P., & Tzetzis, D. (2020). Aerodynamic based shape optimization using CFD: A training case study. *Academic Journal of Manufacturing Engineering*, 18(1), 21–30.

[18] Phadke, G. (2008). *Reduction of Sink Marks in Wire Insert Molded Parts.* Clemson, South Carolina, USA: Clemson University.

[19] Mathivanan, D., Nouby, M., & Vidhya, R. (2010). Minimization of sink mark defects in injection molding process – Taguchi approach. *International Journal of Engineering, Science and Technology*, 2(2), 3–22.

Chapter 4

The Design of Human Robotic Interaction System

Apostolos Tsagaris,[1,*] Vasilis Samaras,[1] Athanasios Manavis,[1] Panagiotis Kyratsis[2]

[1] Department of Industrial Engineering and Management, International Hellenic University, Sindos, Thessaloniki, Greece

[2] Department of Product and Systems Design Engineering, University of Western Macedonia, Kila Kozani, Greece

Abstract

The present chapter includes the study and design of an offline programming methodology for a robotic arm with 5 degrees of freedom. After the design of all the parts of a SCORBOT-ER-V-type robotic arm in exact dimensions, its kinematic model was solved and a custom application was developed in a MATLAB™ programming environment for its complete control. A suitable interface was designed to handle it, and the movement and programming functions were tested in a virtual environment. The application showed the advantages of offline programming but also the integration of different applications from different manufacturers. Thus there is no absolute dependence on the respective manufacturer, but the individual parts of a wider mechatronic or robotic system can be completed with applications of wide and common use. This chapter develops an offline programming and simulation software for industrial robots that integrate the functions of analysis, data processing, 3D simulation, trajectory planning, and program extraction.

Keywords: interface design, offline programming, robotic system, system integration

Introduction

The continuous development of robotics in all the research fields and its evolution in the existing ones are indisputable facts. The advantages are more and more obvious in their application, and their capabilities are still being enriched (Kyratsis, 2020; Kyratsis et al., 2020). However, the desired actions are becoming more demanding, and the evolution of robotic arms is not limited to their material part. Programming and development of appropriate software are necessary for robotic arms to perform the desired tasks. In addition, human presence is not necessary for any setting of the software, as it is possible to do it remotely, as long as there is a connection to a network. With the continued widespread application of industrial robots,

* Corresponding Author Email: tsagaris@ihu.gr

offline programming has gradually become the main mode of industrial robot programming. Thus there is the possibility of imparting higher intelligence to robotic systems (Tsagaris et al., 2012; Tsagaris et al., 2018).

Robots are widely used in modern industrial production processes, and their efficiency and economy are largely determined by their programming methods. When designing the work of industrial robots, there are two methods that can be used: teaching-playing and offline programming. The first robots were mainly used in mass production, such as spot welding on automatic production lines. The tasks were simple and invariant, and the robot's task design could be completed by manual instruction. With the application of robots in small and medium-sized production batches, the complexity of the tasks to be completed increased, the product life cycle became shorter, and production task changes were accelerated. One of the effective ways to solve this problem is to use offline programming to separate the robot from the environment-dependent programming, thereby improving the use efficiency and automation level of the production process and reducing the cost. The offline robot programming system uses the achievements of computer graphics to create a model of the robot and its environment and uses the robot language and related algorithms to perform trajectory planning under offline conditions through the control and operation of graphics (Mitsi et al., 2004). Offline robot programming systems have proven to be a powerful tool to increase safety, reduce robot downtime, and reduce costs (Shangyang et al., 2000).

In the present chapter, a SCORBOT-ER-V-type robotic arm with 5 degrees of freedom (DOF) is used as a research tool. The user needs are analyzed and an offline programming simulation system with satisfactory performance is designed and completed. The main parts of the methodology include the design of the robotic arm in a computer-aided design (CAD) environment, its kinematic analysis with the help of kinematic equations, the design of the control interface, and the development of the control interface (Figure 4.1).

The automation control of a robotic system is achieved by the programming capability and process of its microcontroller. One of the most important advantages of a robotic system is its reprogrammability (Guhl, 2019). With this feature, the user can adjust the operation of the

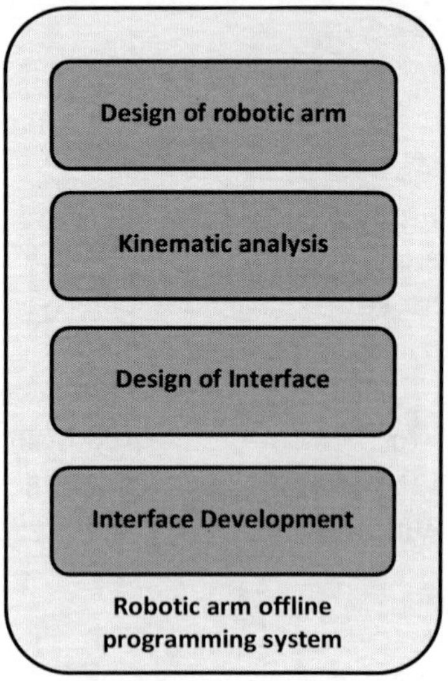

Figure 4.1 Methodology for offline programming system

robot according to the new parameters that will have arisen and in addition without having to purchase a new robotic system that offers the desired operation. Programming in this case is distinguished in two ways: online and offline programming.

During online programming, it is necessary for the robot to stop its operation, in order for the operator to program its new routine, even if it is small modifications to the code. During the programming process, the presence of the robot is usually not necessary. This method tends to be used mainly by users who are not very experienced with programming, but it has disadvantages which are highly predictable. The major disadvantage is the interruption of operation until the programming process is completed. This entails a loss of work and, by extension, a financial loss.

The other way to program the robot is called offline programming. With this method it is possible to reprogram the robot without its long and predictable downtime. During this process, the user performs a simulation of the robotic system in order to see it in operation without disturbing and interrupting the physical robotic system. This is initially achieved by designing the robotic system in a design program (Baizid et al., 2016; Mahmoud and Ilhan, 2016). Then the mathematical model is solved, where the forward or inverse kinematic problem or both will be solved. Then the program is selected which will achieve communication with the graphical environment of the robotic system through its mathematical model. After completing the above steps, the user now has the possibility to carry out several experimental tests until deciding which one is the most suitable (Neto and Mendes, 2013). Finally, the operation of the robotic system is interrupted in order to introduce its new operation routine, and once its programming is complete, it returns to operation mode.

While offline programming is more practical for many reasons, there are cases where its implementation fails. Mainly, in order to carry out the method in question, it is necessary for the user to have programming skills and knowledge. More specifically, since the user is experienced with programming, the choice of offline programming offers more and more important advantages (Carvalho et al., 1998). Some of the most important advantages are presented below:

- Numerical control programs are prepared without stopping the robot, resulting in a reduction in robot shutdown time.
- The programmer is removed from potentially dangerous environments, as most of the program development is performed away from the robot.
- Because offline programming includes process simulation there is an increased possibility of optimizing the layout of the workspace and the work performed by the robot.
- New programs can include previously developed routines.
- Changes to the program can be completed very quickly by replacing only the necessary parts of the program.
- Information from the work environment can be integrated (through design programs) into the program in order to increase the accuracy of the robot process.

In the modeling of a robotic arm, its position analysis and its kinematic state play a very important role. As the robot's DOF increase, finding a solution through inverse kinematics is a difficult process. The conventional methods for calculating the inverse kinematics of any robotic manipulator are geometric (Featherstone, 1983), algebraic (Manocha and Canny, 1994), or iterative (Korein and Balder, 1982) methods. All have advantages and disadvantages and are applied according to the circumstances. Methods incorporating artificial intelligence techniques such as neural networks have also been proposed over time (Lee et al., 1993). It offers the feature of using the fault tolerance and high-speed advantages of neural networks for the problem of solving the inverse kinematic model (Chaudhary et al., 2012). Offline programming can be divided into four levels depending on the level of control at which the programmer defines the tool motion: joint level, robot endpoint level, object level, and task level (Gini, 1987).

At the joint level to complete the required motion, each joint of the robot is programmed individually in the common coordinate space. Since a given work task is mostly described in a Cartesian coordinate space, programming in this case is quite difficult. Even if simple tasks are completed, motion synthesis cannot be completed easily, and the entire programming process is difficult and inefficient. Programming of this kind uses the straight kinematic model of the robotic system.

At the end-effector level for a given task, the trajectory and points in Cartesian coordinate space are programmed using mathematical methods to calculate the movements of each joint according to the position of the manipulator's end-effector. Programming of this kind uses the inverse kinematic model of the robotic system. In both the above cases the programming focuses on the movement of the robot and usually consists of a series of commands that make the final robot move from one position to another, so it is also called action-level programming.

At the object level one does not need to know the specific information of the position of the operator end-operator (the Cartesian coordinate value of the end-operator). The system automatically calculates the coordinate value of the end effector and completes the predetermined work task according to the Cartesian coordinate value of the object. This means that there is a general model of the robotic cell from which the information needed to determine the end manipulator's posture is extracted and used to control the motion.

Finally, the job level is the highest level of robot programming. In task-level programming, the programmer directly issues commands to the robot to perform a specific task (such as welding an object) and through an application of artificial intelligence technologies allows the robot to automatically complete the specified task. This requires not only general model data containing the robot's working environment but also knowledge and intelligent algorithms of the applied process (Carvalho et al., 1998).

Incorporating intelligence techniques into robotic systems is an ever-evolving trend. The capabilities and characteristics of autonomous learning, autonomous decision-making, active interaction, and situation awareness possessed by intelligent systems have brought a large number of new research proposals to the study of human–intelligent system interaction. When the system can act autonomously to some extent, the traditional "stimulus-response" human–computer relationship model changes and the era of human–computer interaction evolves into the era of human–computer integration, so the human–computer behavior model must be considered as a whole (Farooq and Grudin, 2016). Driven by artificial intelligence technology, an enterprise-style relationship of "Human–Intelligent System Collaboration" has begun to emerge. Human–computer collaboration is not a replacement for human–computer interaction, but an extension and development of the concept and its research field. The complexity and uncertainty of widely used intelligent systems can exceed the expectations of the creator, produce unpredictable behaviors, and cause corresponding negative consequences for people and society. Therefore, intelligent human–machine collaboration must be considered from a holistic perspective. It is necessary to start from the technical characteristics of intelligent systems to explore the possibility of intelligent human–machine collaboration in order to address the problems that exist in human–machine collaboration from the perspective of experience, according to human and social challenges (Peeteres et al., 2020).

Kinematic Analysis

Kinematic analysis of a robotic arm is called the analytical study of its movement, without including forces and moments. During kinematic analysis, one encounters two types of kinematics problems which are distinguished into direct and inverse kinematics problems. In the direct kinematics problem, a command is given directly to the individual joints so that the end effector tool moves in the *X,Y,Z* coordinates. Therefore the movement performed by the joint is known and the problem to be studied is to find the coordinates of the end effector.

In the inverse kinematic problem, however, the opposite happens. The user selects a point in *X,Y,Z* coordinates, and the central processing unit calculates the movement of each joint needed for the end effector tool to reach the desired point. In simpler words, the desired position is known and the problem studies the movement and position of each joint in order to achieve it.

According to the above, the definition of kinematic analysis is the transformation of a Cartesian system into a modular one and vice versa. In general there are several methods by which one can calculate the kinematic analysis. The Denavit-Hartenberg (DH) method is the most widespread way to achieve the kinematic analysis of the robotic arm which is used in this work (Denavit and Hartenberg, 1955). The robotic arm used is the SCORBOT-ER V plus. It is an articulated robotic arm, which performs its movements through five rotating joints. The DOF of the robot is 5, and if we include the opening and closing of the gripper, then one more DOF is added and they become 6 (Figure 4.2).

The SCORBOT-ER V plus robotic system has base, body, arm, forearm, and grip joints. These joints move respectively through the joint of the base, shoulder, elbow, and wrist which

Figure 4.2 Structure of SCORBOT-ER V plus

changes the inclination and rotation of the grip. The SCORBOT-ER V plus work envelope defines the total space in which the final action tool can work.

The movement of the joints and the gripper is performed by the DC servomotors (Figure 4.3), where the rotation is determined by the polarity of the operating voltage applied to them: when a positive voltage is applied, the servomotor rotates in one direction, while when a negative voltage is applied, it rotates to the reverse.

The position and movement of each axis is measured by an electro-optical encoder located between the gaps of the motor, which drives the axis (Figure 4.4). As the spindle moves, the encoder produces a series of alternating high and low electrical signals. The sequence of signals defines the direction of movement.

An additional five micro-switches are fitted to the frame of the robotic arm. When the robotic arm defines the position in which the micro-switch is at rest, it defines the position known as the rest position (Figure 4.5).

A variety of means are used to transmit the link motion of the SCORBOT-ER V plus. The base and shoulder shafts are driven by rotating gears, while the elbow shaft is driven by pulleys and timing belts. The rotary movement and tilt displacement of the wrist is carried out by a complex system of pulleys, timing belts, and a differential bevel gear unit. Finally, a lead screw transmits the opening and closing movement of the gripper.

The SCORBOT-ER V plus usually has a built-in gripper with built-in rubber pads. These pads can be removed and replaced with absorbent pads. Three bevel gears form a differential gear system, which drives the wrist joint. When motors 4 and 5 rotate in opposite directions, the tilt of the wrist is shifted, while when the same motors rotate in the same direction, then the

Figure 4.3 SCORBOT-ER V plus servomotors

Figure 4.4 SCORBOT-ER V plus encoder

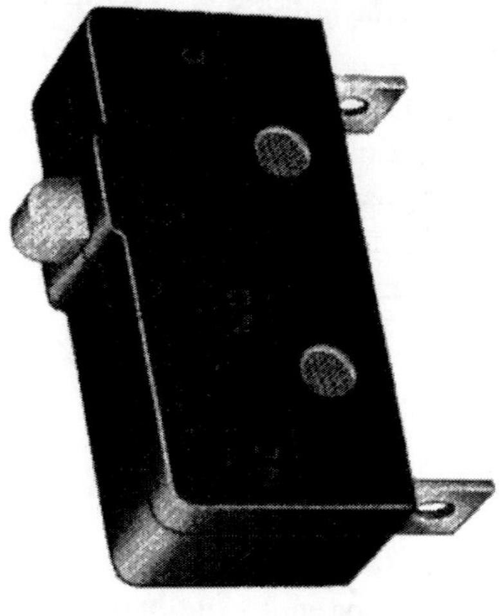

Figure 4.5 Micro-switch – SCORBOT-ER V plus

Figure 4.6 End effector

wrist rotates and by extension the grip. Finally, a pair of guides adapted to motor 6 force the gripper to open and close (Figure 4.6).

Robot Design and Modeling

The design of the individual parts of the robotic system was done with the help of appropriate 3D design software. The software enables the integration of various parts in order to show movement and interaction with each other. It is a CAD program and also offers computer-aided engineering capability. It was used to design SCORBOT-ER V plus connectors and then assemble them to simulate its operation in a graphical environment.

The design of the SCORBOT-ER V plus required the examination and measurement of an identical robotic arm with its actual dimensions. The design and graphical modeling were performed approximately since the purpose of this work is to solve an inverse kinematic problem and offline programming. The parameters taken into account are the following:

- distances between joints
- dimensions of links
- joint movement parameterization
- approximate form of SCORBOT-ER V plus

The first step with which the SCORBOT-ER V plus design was carried out was the measurement of the dimensions of each link separately and then the corresponding design of these in the design program (Figure 4.7).

At each link a coordinate system is also observed which has been chosen based on the DH method (Denavit and Hartenberg, 1955). The above links are then assembled with the assembly option provided by the design program, where the position and movement parameters of the parts between them are determined in order to simulate the robotic system (Figure 4.8).

Above is the complete, unified model with the coordinate systems determined by the DH method. In the kinematic analysis through the DH method, the coordinate frame placed on each joint of the robotic arm must first be defined to derive the four DH parameters. Below is the schematic performance of robotic system coordinate frames (Figure 4.9).

The aim is to calculate the reverse kinematic model to get the information for the angle of the joint according to the movements of the x, y, z position of the end-effector (1).

$$(X, Y, Z) \rightarrow (\Theta 1, \Theta 2, \Theta 3, \Theta 4, \Theta 5) \quad (4.1)$$

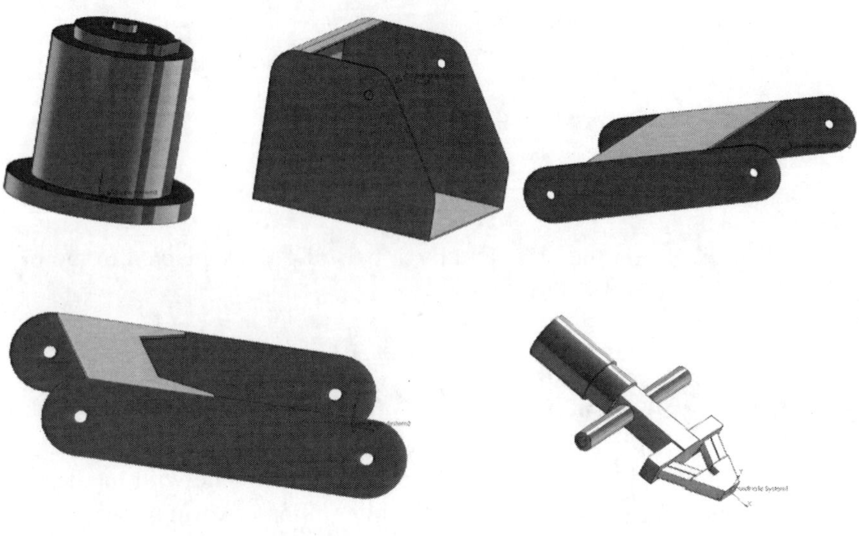

Figure 4.7 Part of robotic system design in 3D modeling

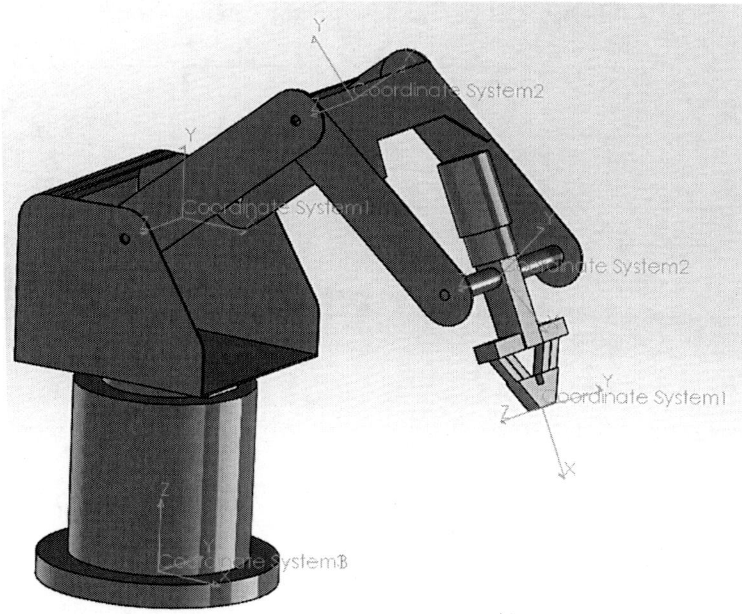

Figure 4.8 The assembly of robotic system

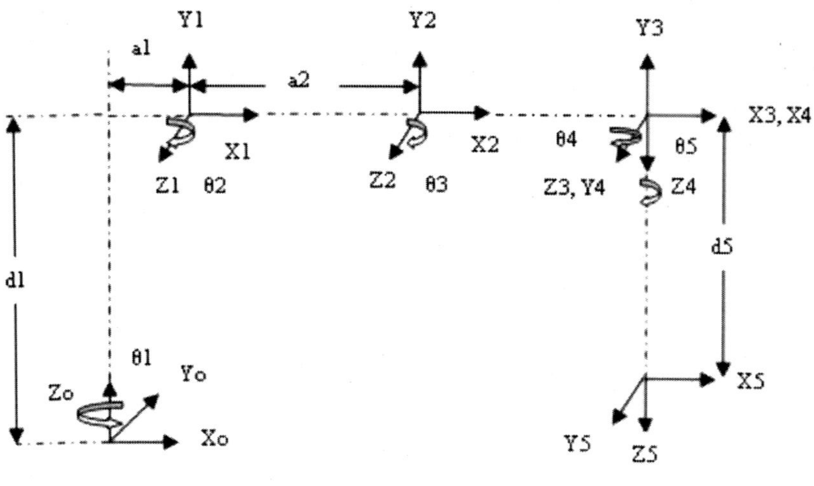

Figure 4.9 Frames for DH calculation

Table 4.1 shows the DH parameters corresponding to each joint, and compared to Figure 4.9, the correlation of each coordinate system and each parameter can be found (Chaudhary and Prasad, 2011).

According to DH methodology, the transformation matrix for joint "*i*" to joint "*I* + 1" is given by (2):

$$^{i-1}T_i = \begin{bmatrix} \cos\theta i & -\sin\theta i\cos\alpha i & \sin\theta i\sin\alpha i & \alpha i\cos\theta i \\ \sin\theta i & \cos\theta i\cos\alpha i & -\cos\theta i\sin\alpha i & \alpha i\sin\theta i \\ 0 & \sin\alpha i & \cos\alpha i & di \\ 0 & 0 & 0 & 1 \end{bmatrix} \quad (4.2)$$

Table 4.1 DH parameters for the robotic system

Joint i	α_i (Degrees°)	a_i (mm)	d_i (*mm*)	Θ_i (Degrees°)	Range
1	90°	101.25	334.25	38.15°	−155 ως +155
2	0°	220	0	−30°	−35 ως +130
3	0°	220	0	45°	−130 ως +130
4	90°	0	0	−63.54°	−130 ως +130
5	0°	0	137.35	0°	−570 ως +570

Multiplying all the ${}^{i-1}T_i$ for $i = 1$ to 5 (${}^0T_1$, ${}^1T_2$, ${}^2T_3$, ${}^3T_4$, ${}^4T_5$), the result is the matrix ${}^0T_5$.
For simplicity reasons is accepted that (3):

$$C_i = \cos\theta_i,\ S_i = \sin\theta_i,\ C_{ijk} = \cos(\theta_i + \theta_j + \theta_k),\ S_{ijk} = \sin(\theta_i + \theta_j + \theta_k) \tag{4.3}$$

$$ {}^0T_5 = \begin{bmatrix} C_1C_{234}C_5 + S_1S_5 & C_1C_{234}S_5 + S_1C_5 & C_1S_{234} & C_1\left(d_5S_{234} + a_3C_{23} + a_2C_2 + a_1\right) \\ S_1C_{234}C_5 - C_1S_5 & -S_1C_{234}S_5 - C_1C_5 & S_1S_{234} & S_1\left(d_5C_{234} + a_3S_{23} + a_2C_2 + a_1\right) \\ S_{234}C_5 & -S_{234}S_5 & -C_{234} & C_1\left(d_5 + a_3C_{23} + a_2C_2 + a_1\right) \\ 0 & 0 & 0 & 1 \end{bmatrix} \tag{4.4}$$

T_e describes the transformation matrix of the end-effector (5).

$$T_e = \begin{bmatrix} n_x & o_x & a_x & p_x \\ n_y & o_y & a_y & p_y \\ n_z & o_z & a_z & p_z \\ 0 & 0 & 0 & 1 \end{bmatrix} \tag{4.5}$$

$$T_e = {}^0T_1 \cdot {}^1T_2 \cdot {}^2T_3 \cdot {}^3T_4 \cdot {}^4T_5 \tag{4.6}$$

$$({}^0T_1)^{-1} \cdot T_e = {}^1T_2 \cdot {}^2T_3 \cdot {}^3T_4 \cdot {}^4T_5 \tag{4.7}$$

$${}^4T_5 = ({}^3T_4)^{-1} \cdot ({}^2T_3)^{-1} \cdot ({}^1T_2)^{-1} \cdot ({}^0T_1)^{-1} \cdot T_e \tag{4.8}$$

From (4), (5), and (6)

$$P_x = C_1 \cdot (S_{234} \cdot d_5 + a_3 \cdot C_{23} + a_2 \cdot C_2 + a_1) \tag{4.9}$$

$$P_y = S_1 \cdot (S_{234} \cdot d_5 + a_3 \cdot C_{23} + a_2 \cdot C_2 + a_1) \tag{4.10}$$

From (9) and (10):

$$\theta_1 = \tan^{-1}\frac{p_y}{p_x} \tag{4.11}$$

From (4), (5), and (6)

$$C_1 \cdot p_x + S_1 \cdot p_y = S_{234} \cdot d_5 + a_3 \cdot C_{23} + a_2 \cdot C_2 \tag{4.12}$$

$$P_x = -C_{234} \cdot d_5 + a_3 \cdot S_{23} + a_2 \cdot S_2 \tag{4.13}$$

$$C_3 = \frac{\left(C_1 \cdot p_x + S_1 \cdot p_y - S_{234} \cdot d_5\right)^2 + (p_z + C_{234} \cdot d_5)^2 - (a_2)^2 - \left(a_3\right)^2}{2 \cdot a_2 \cdot a_3} \tag{4.14}$$

$$S_3 = \pm\sqrt{1 + C_3^2} \tag{4.15}$$

From (14) and (15):

$$\theta_3 = \tan^{-1}\left(\frac{S_3}{C_3}\right) \tag{4.16}$$

From (12) and (13):

$$C_2 = \frac{\left(C_1 \cdot p_x + S_1 \cdot p_y - S_{234} \cdot d_5\right) \cdot \left(a_3 \cdot C_3 + a_2\right) + \left(p_z + C_{234} \cdot d_5\right) \cdot S_3 \cdot a_3}{\left(a_3 \cdot C_3 + a_2\right)^2 + S_3^2 \cdot a_3^2} \tag{4.17}$$

$$S_2 = \frac{\left(C_1 \cdot p_x + S_1 \cdot p_y - S_{234} \cdot d_5\right) \cdot \left(a_3 \cdot S_3\right) + \left(p_y + C_{234} \cdot d_5\right) \cdot \left(a_3 \cdot C_3 + a_2\right)}{\left(a_3 \cdot C_3 + a_2\right)^2 + S_3^2 \cdot a_3^2} \tag{4.18}$$

From (17) and (18):

$$\theta_2 = \tan^{-1}\left(\frac{S_2}{C_2}\right) \tag{4.19}$$

$$\theta_{234} = -\tan^{-1} \cdot \left(\frac{C_1 \cdot a_x + S_1 \cdot a_y}{a_z}\right) \tag{4.20}$$

$$\theta_4 = \theta_{234} - \theta_2 - \theta_3 \tag{4.21}$$

Finally according to Equations (11), (19), (16), and (21) the parameters θ_1, θ_2, θ_3, θ_4, and θ_5 are defined and the reverse kinematic model solved.

Design of User Interface

The design of the interface is completed through the MATLAB™ application. It is a program that provides a full set of applications through a numerical computing environment and a fourth-generation programming language and offers the possibility of entering programming code through a script. The name comes from the words Matrix Laboratory, and it can prove to be highly compatible with external programs, such as some design programs like Solidworks™ and AutoCAD™. It is quite popular and used by millions of scientists and engineers because of the capabilities it offers on its own and with additional applications.

The first step was to calculate the matrices during the kinematic analysis to derive the ${}_{0}T^{5}$ matrix. The next step was the transfer of the robotic system to Simulink™ of MATLAB™. In order to carry out this process an additional application called Simscape™ Multibody™ Link was used. The choice of this application was preferred because it provides compatibility between the design program and Simulink™. Thus, during the realization of the application, the following model emerged in Simulink™ (Figure 4.10).

In the block diagram given in Figure 4.10, the various levels are illustrated, which, as a whole, describe the joints, shapes, and links of the graphic model of the robotic system implemented in the design program.

After being modeled with the help of the programming environment, it is executed and implemented. A window pops up that graphically displays the SCORBOT-ER V plus robotic arm previously designed in the design program. In addition, on the left you can see all the sliders that correspond to each joint and how they have been configured in terms of the range of values they can deliver (Figure 4.11).

The solving of the direct kinematics problem is simulated. Changing the value of the angle of each joint automatically changes the endpoint tool in terms of X, Y, and Z. In other words a value is given in degrees and the coordinates are derived.

The interface was developed through a graphical environment to provide motion control of the graphical model of the robotic arm. For the construction of this graphical environment, the application GUIDE™ (Graphical User Interface Development Environment) was used by the MATLAB™ program, which provides the necessary tools for the construction of a platform

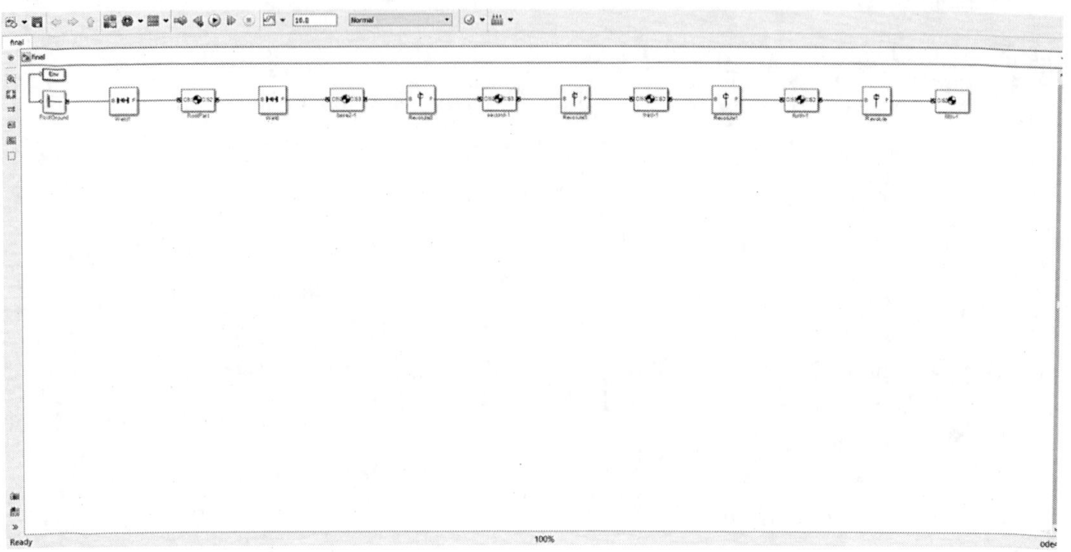

Figure 4.10 Simulink model for the robotic system

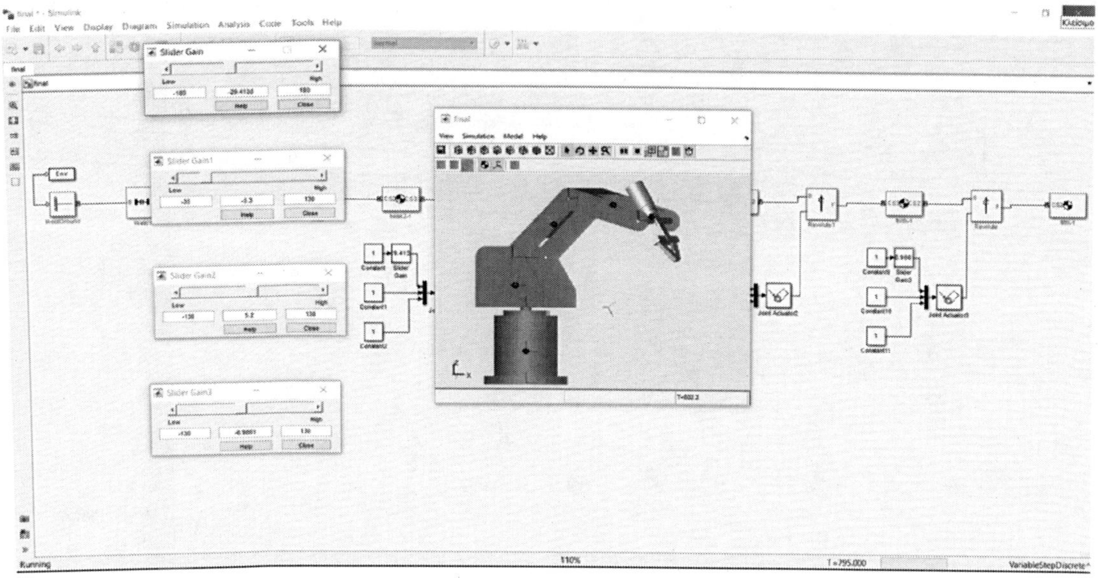

Figure 4.11 3D representation of robotic model in interaction interface

that will interact simultaneously with Simulink™ and with a command code required for this purpose.

As its name describes, GUIDE™ is a graphical user interface (GUI) development environment. The platform with the tools that will be created is a GUI. Through a richly equipped library of tools, the options are given to the user both in handling and entering values and sizes as well as in displaying results, to choose what will provide the greatest convenience and understanding.

Since it is a direct and inverse kinematic problem solution, the GUI should be designed to serve both functions. For the direct kinematic function, it is possible to enter values in each joint, while, respectively, for the inverse kinematic function, in the *X*, *Y*, and *Z* coordinates. Therefore, the GUI has the form shown below (Figure 4.12).

On the left side, under the caption "DIRECT" the direct kinematic mode is checked, while on the right side below the caption "INVERSE" the inverse is checked. For the straight kinematic operation, four linear sliders have been placed which correspond to the four joints.

This completes in software the design of each link, the orientation of each link, the assembly of each link to implement the robotic arm, the transfer of the arm to Simulink™, and the construction of the GUI.

By building and creating the GUI, a set of routines representing the functions of the tools selected and inserted into it is automatically created in the MATLAB™ Editor.

By carrying out the necessary study for the desired function of the GUI in terms of the robotic arm, a set of elements is obtained which are to be inserted into the routines and a series of commands and calculations are executed which will lead to the desired result. The necessary steps to achieve forward and reverse kinematic operation are described below.

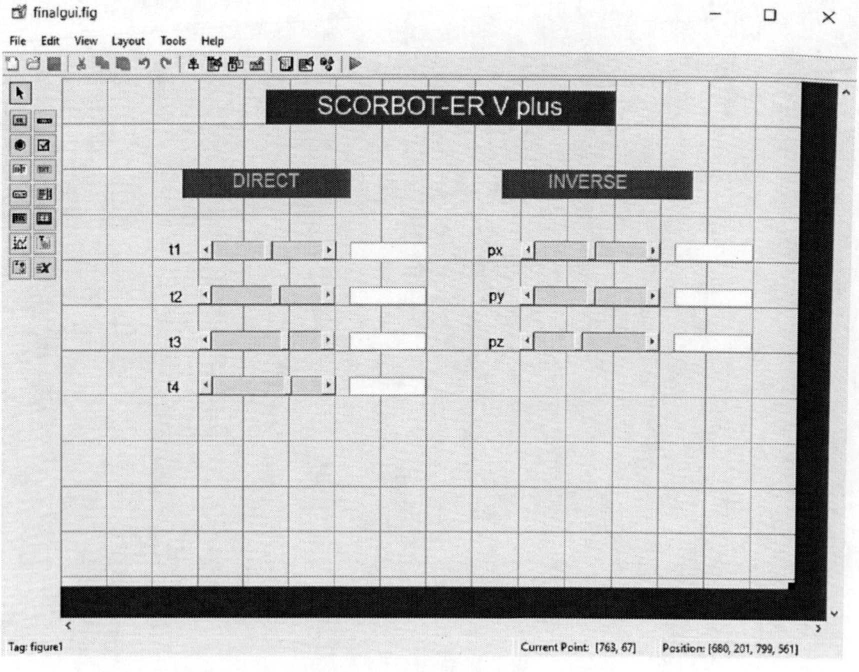

Figure 4.12 Design of GUI for interaction with robot

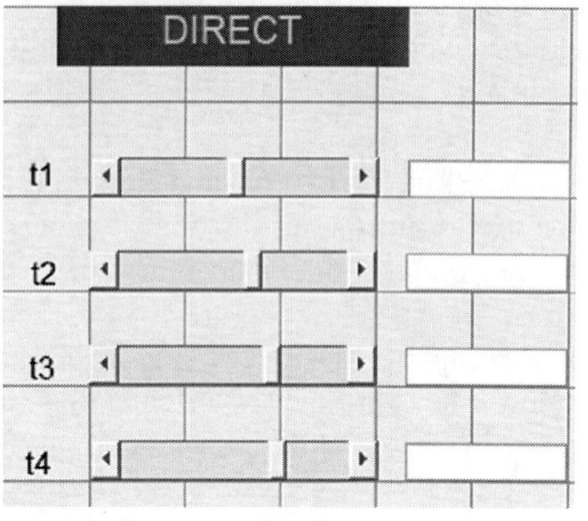

Figure 4.13 GUI design for forward kinematic model

Because the purpose of the work is to study the motion of the end-action tool without additional functions and capabilities, the 2 DOF are removed. The first degree concerns the rotation of the final action tool, and the second the opening-closing of the gripper. So since it is redundant to continue and complete it, for the sake of simplification, the rest of the work will be performed to control a robotic arm with 4 DOF.

Forward Kinematic Model

During the direct kinematic operation of an arm, the values in degrees are entered for each joint separately, and at the output, values for the *X,Y,Z* coordinates are obtained, which correspond to the position of the final action tool. Since it is about the direct kinematic mode, the code will be entered in the routines connected to the sliders under the "DIRECT" label of the GUI (Figure 4.13).

Where "t1", "t2", "t3", and "t4" are the values that will be entered respectively in the joints "base", "shoulder", "elbow", and the "wrist inclination". As you can see, the set of commands is divided into four pairs of lines, where the same function is repeated in each pair, but for a different articulation. In the first line it is shown that the value of each joint is equal to the value received by the corresponding slider during its change. Then, in the second line of each pair, the value corresponding to the joint (code 1) is placed as an indicator to the right of each slider.

```
t1=get(handles.slider1,'value');
set(handles.edit1,'string',num2str(t1));

t2=get(handles.slider2,'value');
set(handles.edit2,'string',num2str(t2));

t3=get(handles.slider3,'value');
set(handles.edit3,'string',num2str(t3));

t4=get(handles.slider4,'value');
set(handles.edit4,'string',num2str(t4));
```

Code 1. Commands to control via GUI the direct kinematic model

Immediately after that, the following set of commands is entered in which, with each new change of each joint, the controllers that were placed in Simulink™ and by extension the joints themselves are automatically changed, thus making the desired movement of the robotic arm in the graphical environment (Code 2).

```
set_param([ModelName '/Slider Gain'],'Gain',num2str(t1))
set_param([ModelName '/Slider Gain1'],'Gain',num2str(t2))
set_param([ModelName '/Slider Gain2'],'Gain',num2str(t3))
set_param([ModelName '/Slider Gain3'],'Gain',num2str(t4))
```

Code 2. Connection with Simulink GUI

Also above, the solution of the direct kinematic problem was carried out, during which a set of equations was obtained. In order to perform the direct kinematic operation, it is sufficient to use the ${}^{0}T_4$ array. For better understanding, the same matrices will be added into the code, the product of which results in ${}^{0}T_4$. The matrices are ${}^{0}T_1$, ${}^{1}T_2$, ${}^{2}T_3$, and ${}^{3}T_4$.

So, then, the following commands-actions are introduced where the calculations are performed again for the new data (Code 3).

```
T1=[cosd(t1) 0 sind(t1) 10.87*cosd(t1);sind(t1) 0 -cosd(t1) 10.87*sind(t1);0 1 0 352; 0 0 0 1];
T2=[cosd(t2) -sind(t2) 0 236*cosd(t2); sind(t2) cosd(t2) 0 236*sind(t2);0 0 1 0;0 0 0 1];
T3=[cosd(t3) -sind(t3) 0 220*cosd(t3); sind(t3) cosd(t3) 0 220*sind(t3);0 0 1 0;0 0 0 1];
T4=[cosd(t4) -sind(t4) 0 131*cosd(t4); sind(t4) cosd(t4) 0 131*sind(t4);0 0 1 0;0 0 0 1];

T=T1*T2*T3*T4;
```

Code 3. New position calculation

Since the desired table ${}^{0}T_4$ has been generated, all that remains is to take the first three elements of its last column to obtain the coordinates of the endpoint tool *X*, *Y*, and *Z*, respectively. This is accomplished with the following set of commands, where "px", "py", and "pz" are the corresponding values of the coordinates "*X*", "*Y*" and "*Z*", and the numbers in parentheses are indicators of the positions of the table where the first one symbolizes the line and the second the column (Code 4).

```
px=T(1,4);
py=T(2,4);
pz=T(3,4);
```

Code 4. End-effector position calculation

Finally follows the following set of commands whose function is to configure the sliders and indicators in the GUI that are under the caption "INVERSE" which are associated with the *X*,*Y*,*Z* coordinates of the final action tool. As observed, three pairs of commands follow, where each pair addresses respectively the slider and the indication of each coordinate axis. So, with each change of the value of any joint, the following set of commands make sure that the indicators and sliders are configured appropriately (Code 5).

```
set(handles.slider5,'value',px);
set(handles.edit5,'string',num2str(px));

set(handles.slider6,'value',py);
set(handles.edit6,'string',num2str(py));

set(handles.slider7,'value',pz);
set(handles.edit7,'string',num2str(pz));
```

Code 5. Connection with Simulink™ GUI for kinematic model

After each command and the purpose of their operation have been described in detail, below is illustrated the set of all the above as they have been written in the routine of the first slider (slider1) of the editor in MATLAB™ (Code 6).

```
finalgui.m × | final.m × | +

% --- Executes on slider movement.
function slider1_Callback(hObject, eventdata, handles)
ModelName='scorbot';
global var;

t1=get(handles.slider1,'value');
set(handles.edit1,'string',num2str(t1));

t2=get(handles.slider2,'value');
set(handles.edit2,'string',num2str(t2));

t3=get(handles.slider3,'value');
set(handles.edit3,'string',num2str(t3));

t4=get(handles.slider4,'value');
set(handles.edit4,'string',num2str(t4));

set_param([ModelName '/Slider Gain'],'Gain',num2str(t1))
set_param([ModelName '/Slider Gain1'],'Gain',num2str(t2))
set_param([ModelName '/Slider Gain2'],'Gain',num2str(t3))
set_param([ModelName '/Slider Gain3'],'Gain',num2str(t4))

T1=[cosd(t1) 0 sind(t1) 10.87*cosd(t1);sind(t1) 0 -cosd(t1) 10.87*sind(t1);0 1 0 352; 0 0 0 1];
T2=[cosd(t2) -sind(t2) 0 236*cosd(t2); sind(t2) cosd(t2) 0 236*sind(t2);0 0 1 0;0 0 0 1];
T3=[cosd(t3) -sind(t3) 0 220*cosd(t3); sind(t3) cosd(t3) 0 220*sind(t3);0 0 1 0;0 0 0 1];
T4=[cosd(t4) -sind(t4) 0 131*cosd(t4); sind(t4) cosd(t4) 0 131*sind(t4);0 0 1 0;0 0 0 1];
T=T1*T2*T3*T4;
px=T(1,4);
py=T(2,4);
pz=T(3,4);

set(handles.slider5,'value',px);
set(handles.edit5,'string',num2str(px));

set(handles.slider6,'value',py);
set(handles.edit6,'string',num2str(py));

set(handles.slider7,'value',pz);
set(handles.edit7,'string',num2str(pz));
```

Code 6. MATLAB™ code from interface in forward kinematic model

The above set of commands remains the same in the routines for all the sliders of the "base", "shoulder", "elbow", and "wrist tilt" joints ("slider1", "slider2", "slider3", and "slider4").

Reverse Kinematic Model

There are various applications where the collective movement of the required joints is required for the endpoint tool to come to a very precise and specific *X,Y,Z* coordinate position. In the inverse kinematic problem, exactly the process by which it will be solved is studied, so that in the end a central processing unit is in place to handle a robotic system.

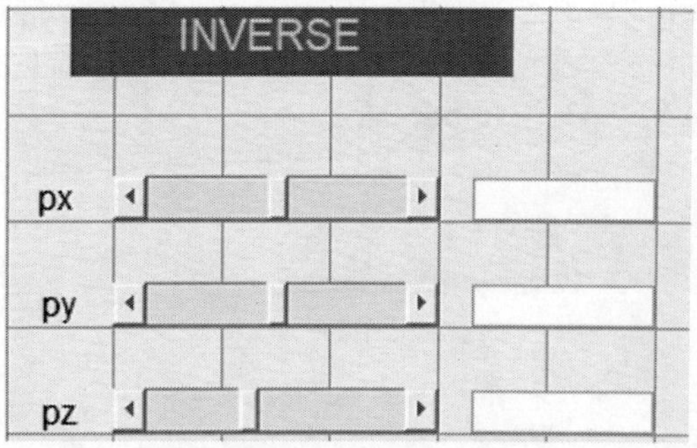

Figure 4.14 GUI design for reverse kinematic model

The DH method was used to solve the inverse kinematic problem. In this analysis, some equations have been derived, which are used in the form of commands and actions in the MATLAB^{TM} Editor to achieve the inverse kinematic function of the robotic system.

As mentioned above in the implementation of the direct kinematic function, the solution was simplified for a robotic system with 4 DOF, because neither the rotation of the end-action tool nor the opening-closing of the gripper is necessary to achieve the purpose of this work.

According to the above, it is clear that the input of the system in this mode will be the three values of the coordinates *X,Y,Z* which will be given directly to the GUI, to the sliders that are under the legend "INVERSE" (Figure 4.14).

Then, the following set of commands is initially introduced into the routine of the first slider labeled "px" (slider5), where it is presented in the form of three pairs. The first line of each pair addresses the value that will be given by the slider, and the second the configuration of the indicator, which will display the exact value of the slider. Finally, the three pairs of these commands appear, each of which corresponds to one of the three coordinate axes (Code 7).

```
px=get(handles.slider5,'value');
set(handles.edit5,'string',num2str(px));

py=get(handles.slider6,'value');
set(handles.edit6,'string',num2str(py));

pz=get(handles.slider7,'value');
set(handles.edit7,'string',num2str(pz));
```

Code 7. Commands to control via GUI the reverse kinematic model

During the DH analysis, the equations were derived that calculate the value of each joint in order for the endpoint tool to be in a certain position. Taking into account minus 2 DOF, the following simplified form is obtained, where the calculation of each of the four joints is observed only with the input of the three coordinates (Code 8).

```
t1=atan2d(py,px);

nx=px*cosd(t1)+py*sind(t1)-10.87-131;
ny=352-pz;
M=(nx*nx+ny*ny-236*236-220*220)/(2*236*220);
N=sqrt(1-M*M);
t3=atan2d(N,M);

A=ny*(220*cosd(t3)+90)-220*sind(t3)*nx;
B=nx*(220*cosd(t3)+90)+220*sind(t3)*ny;
t2=atan2d(A,B);

t4=-t2-t3;
```

Code 8. Commands to calculate the angle of joints

Four pairs of commands are observed, where in each pair the first row updates the corresponding indicator in the GUI depending on what value each joint has. The second line of each pair appropriately configures the corresponding "Slider Gain" in Simulink, which in turn changes the corresponding joint of the robotic arm in the graphical environment (Code 9).

```
set(handles.edit1,'string',t1);
set_param([ModelName '/Slider Gain'],'Gain',num2str(t1))

set(handles.edit2,'string',t2);
set_param([ModelName '/Slider Gain1'],'Gain',num2str(t2))

set(handles.edit3,'string',t3);
set_param([ModelName '/Slider Gain2'],'Gain',num2str(t3))

set(handles.edit4,'string',t4);
set_param([ModelName '/Slider Gain3'],'Gain',num2str(t4))
```

Code 9. Connection with Simulink™ GUI for reverse kinematic model

The aforementioned command sections are compiled and illustrated below into a single set of commands. With the following code, the inverse kinematic operation of the robotic arm is achieved in the Simulink™ GUI. The same code has been placed in the slider routines for the *X,Y,Z* coordinates (slider5, slider6, slider7) (Code 10).

```
function slider5_Callback(hObject, eventdata, handles)

ModelName = 'scorbot';
global var;

px=get(handles.slider5,'value');
set(handles.edit5,'string',num2str(px));

py=get(handles.slider6,'value');
set(handles.edit6,'string',num2str(py));

pz=get(handles.slider7,'value');
set(handles.edit7,'string',num2str(pz));


t1=atan2d(py,px);

nx=px*cosd(t1)+py*sind(t1)-10.87-131;
ny=352-pz;
M=(nx*nx+ny*ny-236*236-220*220)/(2*236*220);
N=sqrt(1-M*M);
t3=atan2d(N,M);

A=ny*(220*cosd(t3)+90)-220*sind(t3)*nx;
B=nx*(220*cosd(t3)+90)+220*sind(t3)*ny;
t2=atan2d(A,B);

t4=-t2-t3;

guidata(hObject,handles);
set(handles.edit1,'string',t1);
set_param([ModelName '/Slider Gain'],'Gain',num2str(t1))

set(handles.edit2,'string',t2);
set_param([ModelName '/Slider Gain1'],'Gain',num2str(t2))

set(handles.edit3,'string',t3);
set_param([ModelName '/Slider Gain2'],'Gain',num2str(t3))

set(handles.edit4,'string',t4);
set_param([ModelName '/Slider Gain3'],'Gain',num2str(t4))
```

Code 10. MATLAB™ code from interface in reverse kinematic model

Application

This section presents the complete operation of the robotic arm in the Simulink graphical environment, where it will be controlled through a GUI. The direct and inverse kinematic functions are presented in turn through examples and their explanation.

The arm is in its initial position since the joint angles are at the zero position. This is set directly by the sliders in the left column labeled "DIRECT". It is observed that the joints "t3" and "t4" are not absolutely zero. This happens due to the execution of the code, where during its operation it is unable to completely zero the variables in question. Thus, it approaches zero as much as possible resulting in a value that offers a negligible deviation for the motion data of the particular robotic arm (Figure 4.15).

Then a random movement of each joint through the corresponding slider is depicted. During this action a series of events are observed. Initially, the slider indicator of each joint

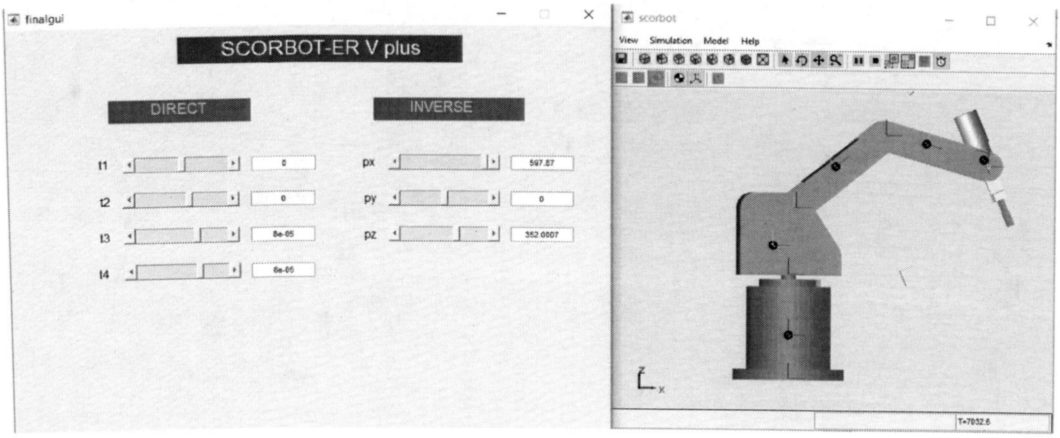

Figure 4.15 Robotic arm representation

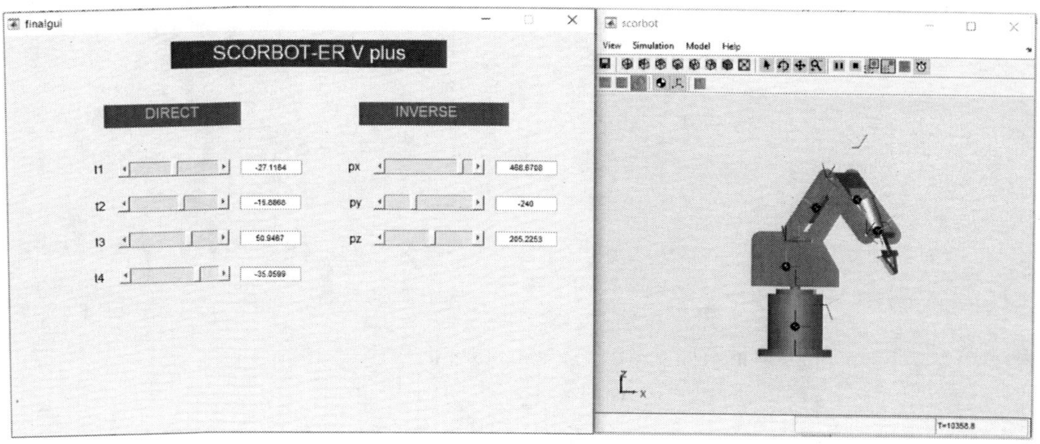

Figure 4.16 Robotic arm moving using forward kinematic values

takes the specific value desired by the user. Then the indications on the coordinate sliders are automatically changed, which constantly depict the exact position of the final action tool in space. Finally, in the graphical environment, it is observed that with the change in the value of each joint, the corresponding joint can be seen moving in real time (Figure 4.16).

When applying inverse kinematics, the movement of the arm is presented with values at the endpoint and changes in the joints of the system. A random position of the end-action tool is set which is shown as lower right. Then a sequence of three examples is gradually presented, in which the final action tool performs a diagonal movement in space (Figure 4.17).

Through these three actions a series of events is observed. First, the value change of the indicator next to each slider that changes is displayed. In addition, with the change of each coordinate slider, the value change of the indications of the joints necessary for the desired position is observed. Finally, every time any coordinate controller is changed, the corresponding movement of the robotic arm is observed in real time.

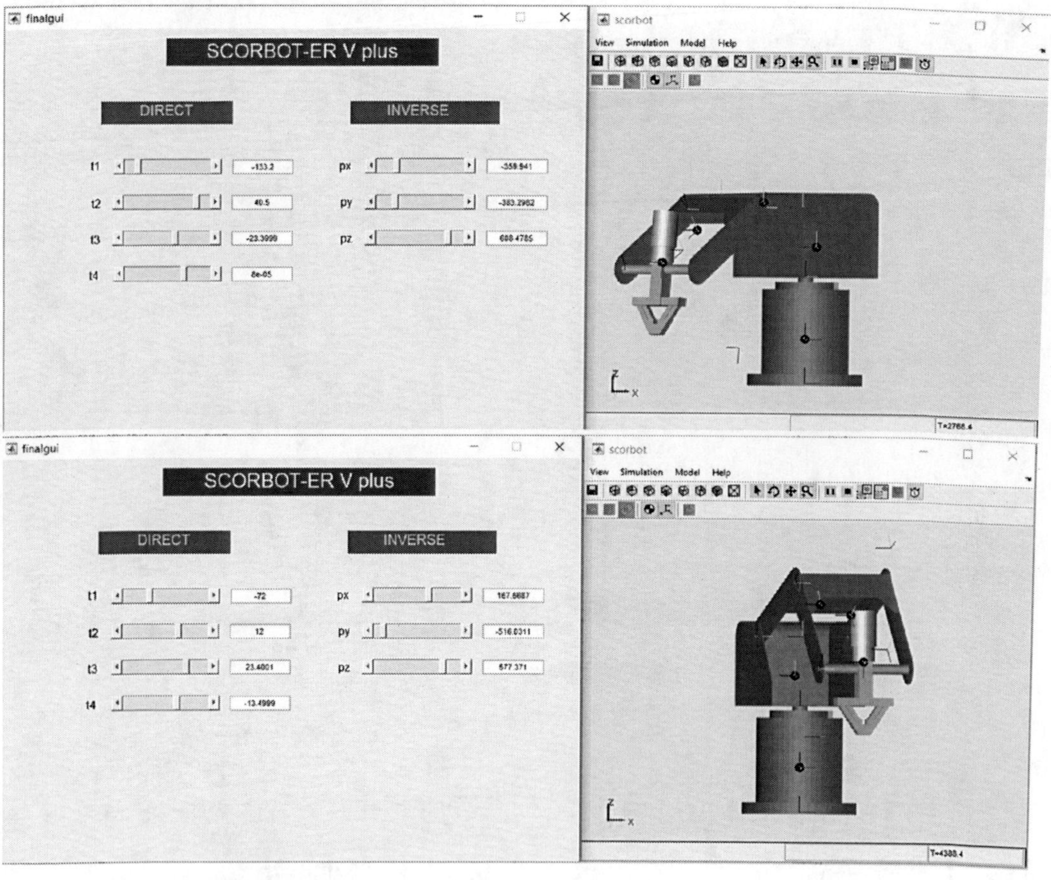

Figure 4.17 Robotic arm moving using reverse kinematic values

Conclusion

It is necessary to transfer the robotic arm to a graphical environment through a useful design program. In the design what should be taken into account are the dimensions of the links, the movements made by the actuators, and the detailed illustration of the end-action tool. Then, using the DH method, the description of the robotic arm is achieved with mathematical equations and vector matrices, thus solving the mathematical kinematic model for forward and reverse operation. In addition, it is important to advance the robotic arm model and kinematic analysis to another program where they will be interconnected in order to achieve the goal of offline programming. With this process, a simulation environment has been created where the user has the ability to make changes and adjustments to a robotic arm from a remote point, thus offering comfort and safety. Also in this way, it is not necessary to interrupt the work of the robotic arm, since there is the possibility of developing the new function in the simulation program, and then the promotion of the new information, directly to its processing unit.

Appropriate design programs and MATLAB™ software were used for the implementation and construction of the above. The design program offers all the necessary tools for the design and implementation of any mechanical model, as well as the precise placement and definition of various coordinate systems. MATLAB™ provides compatibility with the design program through an add-on tool for advancing the robotic arm in Simulink™. Also using GUIDE™ from MATLAB™, it was possible to build a GUI development environment, where by inserting the necessary routines resulting from the kinematic analysis, it was possible to interact and manipulate the robotic arm.

Additionally, this work offers the possibility of its evolution where the SCORBOT-ER V plus robotic arm will perform more complex and complex movements and actions. Suggested actions will be the rotation of the wrist, opening and closing of the gripper, and the introduction of routines for automated repetitive or non-repetitive movement of the robotic arm to fulfill desired processes.

References

Baizid K., Cukovic S., Iqbal J., Yousnadj A., Chellali R., Meddahi A., Devedzic G., Ghionea I. IRoSim: Industrial robotics simulation design planning and optimization platform based on CAD and knowledge ware technologies. *Robotics and Computer-Integrated Manufacturing* (2016) 42: 121–134.

Carvalho G.C., Siqueira M.L., Absi-Alfaro S.C. Off-line programming of flexible welding manufacturing cells. *Journal of Materials Processing Technology* (1998) 78(1–3): 24–28, ISSN 0924-0136.

Chaudhary H., Prasad R. Intelligent inverse kinematic control of scorbot-er v plus robot manipulator. *International Journal of Advances in Engineering & Technology* (2011) 1(5): 158–169, ISSN: 2231-1963.

Chaudhary H., Prasad P., Sukavanum N. Position analysis based approach for trajectory tracking control of scorbot-er-v plus robot manipulator, *International Journal of Advances in Engineering & Technology* (2012) 3(2): 253–264, ISSN: 2231-1963.

Denavit J., Hartenberg R. A kinematic notation for lower-pair mechanisms based on matrices. *Journal of Applied Mechanics, Transactions ASME* (1955) 77: 215–221.

Farooq U., Grudin J. Human-computer integration. *Interactions* (2016) 23(6): 26–32.

Featherstone R. Position and velocity transformation between robot end-effector coordinate and joint angle. *International Journal of Robotics* (1983) 2(2): 35–45.

Gini M. The Future of robot programming. *Robotics* (1987) 5(2): 235–246.

Guhl J., Nikoleizig S., Heimann O., Hügle J., Krüger J. Combining the advantages of on- and offline industrial robot programming. *2019 24th IEEE International Conference on Emerging Technologies and Factory Automation (ETFA)* (2019): 1567–1570, doi:10.1109/ETFA.2019.8869495.

Korein J.U., Balder N.I. Techniques for generating the goal-directed motion of articulated structures. *IEEE Computer Graphics and Application, IEEE* (1982) 2(9): 71–81.

Kyratsis P. Computational design and digital manufacturing applications. *International Journal of Modern Manufacturing Technologies* (2020) 12(1): 82–91.

Kyratsis P., Kakoulis K., Markopoulos A. Advances in CAD/CAM/CAE technologies. *Machines* (2020) 8(1): 13.

Lee D.M.A., Buchal R.O., Elmaraghy W.H., Jack H. Neural networks and the inverse kinematics problem. *Journal of Intelligent Manufacturing* (1993) 4: 43–66.

Mahmoud N., Ilhan E.K. Off-line nominal path generation of 6-DoF robotic manipulator for edge finishing and inspection processes. *International Journal of Advanced Manufacturing Technolog* (2016) 99(1–4): 85–96.

Manocha D., Canny J.F. Efficient inverse kinematics for general 6r manipulators, *IEEE Transaction on Robotics Automation, IEEE* (1994) 10(5): 648–657.

Mitsi S., Bouzakis K.D., Mansour G., Sagris D., Maliaris G. Off-line programming of an industrial robot for manufacturing. *International Journal of Advanced Manufacturing Technolog* (2004) 26(3): 262–267.

Neto P., Mendes N. Direct off-line robot programming via a common CAD package. *Robotics and Autonomous Systems* (2013) 61(8): 896–910.

Peeteres M., Diggelen J.V., Bosch K.V.D. Hybrid collective intelligence in a human: AI society. *AI & Society* (2020) 3: 1–22.

Shangyang L., Shanben Ch., Chengtong Li. Welding robots and their applications. *Machinery Industry Press* (2000): 131–133.

Tsagaris A., Efkolidis N., Stampoulis K., Kyratsis P. Reverse engineering techniques integrated into robotic system for surface reconstruction. *Academic Journal of Manufacturing Engineering* (2012) 10(3): 113–119.

Tsagaris A., Stampoulis K., Kyratsis P. Hand finger gesture modeling in advanced CAD system. *Computer Aided Design and Applications* (2018) 15(3): 281–287.

Chapter 5

Modeling and Simulation of Nonconventional Machining Processes

Daniel Ghiculescu*, Bogdan Cristea, Gabriela Parvu, Cristina Iuga, Mihaela Cirstina

Manufacturing Engineering Department, Polytechnic University of Bucharest, Romania

Abstract

Some nonconventional machining processes are approached concerning modeling and numerical simulation of material removal mechanisms and related processes. From this point of view, different numerical calculation modules are used corresponding to the type of energy used directly for material removal: thermal energy at Electrical Discharge Machining (EDM), Laser Beam Machining (LBM), and Plasma Machining (PM), electrical energy at Electro-Chemical Machining (ECM), as well as Fluid Flow for electrolyte solution supplying at ECM and Microfluidics.

An essential part of a technological system for a hybrid nonconventional process, Ultrasonically aided micro-Electrical Discharge Machining (μEDM+US), namely the feed system is modeled in what concerns the deformation in different cases of functioning along with the determination of its own frequency to avoid the overlapping with the frequency used to command the stepping motor.

Some other thermal processes are modeled like Electron Beam and Ion Beam Machining but from the point of view of tracing electrically charged particles in a magnetic field, in the process of Electrical Discharge Deposition (EDD), in connection with its precision.

Finally, modeling and numerical simulation of microfluidics within the circuits of micro-electro-mechanical systems (MEMS) are approached. These MEMS are achieved by nonconventional technologies and are used in medical fields for determining the number of leucocytes.

Keywords: EDM, μEDM+US, ECM, LBM, BM, EBM, IBM, MEMS.

Modeling of EDM Finishing on Some Advanced Materials

The process of electrical discharge machining using finishing modes was applied to some materials with high characteristics, so-called advanced materials that have multiple applications in such different fields. For instance, CoCr alloys are used in many engineering fields such

---

* Corresponding Author Email: daniel.ghiculescu@upb.ro

as turbomotors, nuclear, biomedical, and dentistry (Vaicelyte et al., 2020). These utilizations are based on their excellent characteristics of corrosion resistance, wear resistance, high temperature resistance, and good biocompatibility. Therefore, CoCr alloys are also widely applied in special fields of medicine like stents (Gherbesi and Natalini, 2020), intervertebral disc replacement, and knee or hip arthroplasty (Louwerens et al., 2020; Liu et al., 2020). These alloys achieve their corrosion resistance by forming chromium-based oxides on the surface, associated with their biocompatibility, becoming very beneficial (Buser et al., 2017).

High-pressure compressors use conventional titanium alloys (α and α-β) in the fields where the working temperature does not go over 5,000 °C (Leyens and Peter, 2003). TiAl-based alloys are especially suitable for applications of low-pressure turbine blades and high-pressure compressor blades (Liu and Liu, 2015). Turbine blades from TiAl alloys equipped wide-body jet airliners.

The main hardening mechanism of cobalt-based alloys is the formation of carbides which disperse in the alloy matrix and precipitate at the granule boundaries. This dispersion has a direct effect on the mechanical strength of the alloy (Mori et al., 2012).

Titanium γ (TiAl) aluminum has the highest specific stiffness compared to all other classes of alloys. In this context, titanium aluminum alloys have high values of modulus of elasticity, compared to conventional titanium alloys and nickel-based alloys (Leyens and Peter, 2003).

Compositions of some CoCr alloys and titanium aluminides (Szkliniarz and Szkliniarz, 2021) that are studied in this chapter are presented in Tables 5.1 and 5.2.

The mechanical properties of some CoCr alloys and titanium aluminides (Szkliniarz and Szkliniarz, 2021) approached in this chapter are shown in Tables 5.3 and 5.4.

Alloys P1 and P3, which contain W, have a slightly higher thermal conductivity compared to alloy P2, as this alloying element has the highest heat transfer coefficient: 170 (W/m K). According to the literature, the specific heat of CoCrMo ternary alloys (similar to alloy 2) is 452 (J/kg grd) (Baron et al., 2015).

Titanium aluminides based on intermetallic compounds γ (TiAl) and α2 (Ti_3Al) have important thermo-physical and mechanical properties, such as high melting point, above 1460°C, low density (3.9–5 g/cm^3, depending on the degree alloy), high stiffness, yield strength and creep

Table 5.1 Chemical composition of CoCr alloys studied

Alloy	Cr (%)	Mo (%)	W (%)	Nb (%)	Si (%)	Mn (%)	Fe (%)	Co (%)
P1, SYSTEM NE	21.0	6.5	6.4	—	0.8	0.65	<0.1	64.4
P2, SYSTEM SOFT	29.5	5.7	—	—	0.95	0.55	0.75	61.8
P3, LUKA CHROM	24.0	3.0	8.0	1.0	1.0	—	—	63.0

Table 5.2 Chemical composition of studied titanium aluminides

Alloy	Alloy element	Ti (%)	Al (%)	Nb (%)	Cr (%)	Zr (%)	Ta (%)	B (%)	α_2 phase (%)
Alloy 1	at %	45	44	4	—	4	1	1	2
Alloy 2	at %	43	47	7	1	—	0.5	—	25

Table 5.3 Mechanical properties of studied CoCr alloys

Alloy	Tensile strength (MPa)	Young Modulus (GPa)	Elastic limit (MPa)	Breaking elongation (%)	Hardness (HV)
P1, SYSTEM NE	850	155	580	3	460
P2, SYSTEM SOFT	447	160	450	15	310
P3, LUKA CHROM	550	125	545	25	185

Table 5.4 Mechanical properties of studied CoCr alloys

Alloy	Tensile strength, at 25°C (MPa)	Tensile strength, at 760°C (MPa)	Breaking elongation, at 25°C (%)
Alloy 1 (samples 1, 2, 3)	620	550	3
Alloy 2 (samples 4, 5)	1,100	620	2-3

resistance, structural stability at high temperatures, and low diffusion coefficient, all being refractory characteristics.

Samples of each alloy presented above were machined by EDM finishing on Romanian EDM machine, ELER 01, using several tool-electrodes with dimensions from Figure 5.1.

Square shape cavities with sides of 5 mm were generated on the samples from studied alloys; the machining data is presented below. The dimensions of the active part of the electrodes, respectively, the size of the cross section (Ds), and the height (hs) were calculated using the relations (Ghiculescu, 2004):

$$D_s = D_p - 2\, s_L \text{ (mm)} \tag{5.1}$$

$$h_s = h_p\,(1+\Theta) + c_s \text{ (mm)} \tag{5.2}$$

where D_p—the size of the part cavity; s_L—side gap (from the ELER 01 user manual); h_p—piece height (mm); Θ—relative wear (from the ELER 01 user manual) (%); c_s—safety factor, c_s=3–5 mm.

The tools were executed on the CNC machining center MAZAK NEXUS 350-II MY. The cutting speed was 40 m/min for rough turning, 50 m/min for finishing turning, and 35 m/min for milling. Turning was performed with a roughing tool (PWLNR 2525 M08—WALTER) and a finishing tool (SDHCR 2525 M11—WALTER) and milling with a Φ6 mm milling cutter (MC111-06.0A4A-WJ30TF—WALTER). The average roughness (Ra) obtained on the frontal surfaces was 0.230 μm and on the side surfaces, 0.325 μm, even better than that specified in the execution draft.

EDM performed on the ELER 01 machine had the following working modes: at preliminary roughing phase, commanded pulses, positive polarity at CoCr alloys and negative at Ti aluminides, current $I = 6A$ (possible maximum current allowed by current density, $J = I/S$, respective 20 A/cm^2, where S-frontal surface), pulse time, $t_i = 95$ μs, pause time, $t_0 = 24$ μs; at final finishing phase, negative polarity, relaxation pulses, $t_i \approx 0{,}8$ μs, given by the capacity steps,

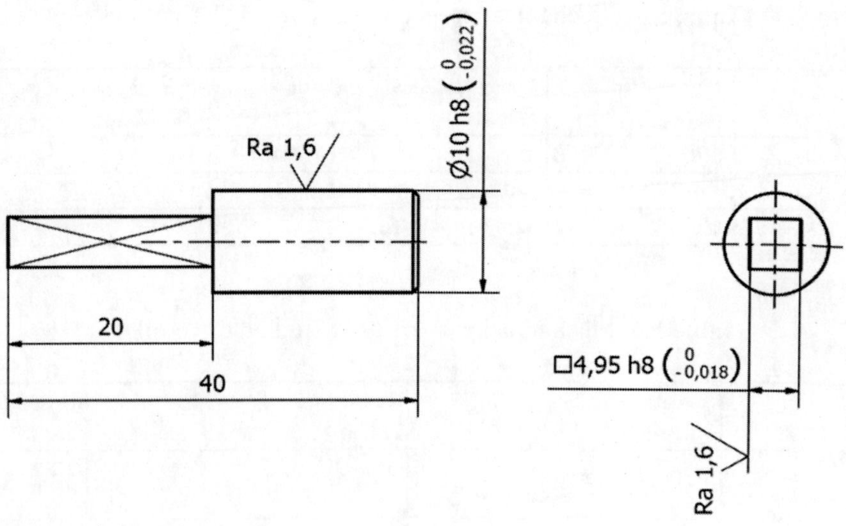

Figure 5.1 Tool-electrode for EDM finishing

C = 10 nF and resistance, R = 0,74 kΩ of finishing EDM generator. At all working modes, the pulsating method was adopted to facilitate the flushing of the working gap: machining time—3s, lifting time—2s.

The experimental data for roughing EDM (R) and finishing (F) are presented in Table 5.5.

According to the results presented above, it can be observed that the machining rate at roughing is much lower at Titanium aluminides than in the case of CoCr-based alloys. At finishing modes, the values of machining rate obtained are relatively close for both main types of materials. This could be explained by the structure of Titanium aluminides due to intense thermal influence of roughing modes, affecting the thermo-physical properties, and consequently the EDM removal mechanism which in case of finishing modes is not so apparent. In terms of machined surface roughness, the values obtained for CoCr-based alloys are lower than in the case of Titanium aluminides, this being influenced by some instability of EDM process (a number of short-circuits occurred between tool and workpiece) noticed during experiments.

Some images of the machined surface provided by Scanning Electron Microscope, QUANTA INSPECT F50 with field emission electron gun with a resolution of 1 nm are presented in Figures 5.2 and 5.3:

The microtopographies are very different for the two types of materials. In the case of CoCr-based alloy, a great quantity of molted material remained resolidified on the machined surface against Ti aluminides, when this phenomenon is not so apparent. The diameters of the craters are generally larger at CoCr machining than in the second case.

The material removal mechanism for a single discharge on the studied materials was numerically modeled and simulated in Comsol Multiphysics™, using the module (physics) of heat transfer in solids, time-dependent variant. A model with pulse time-dependent plasma channel was studied.

In the first stage, the model was parameterized (Figure 5.4).

In the second stage, the geometry of the model was created, using the Comsol™ built-in tools, as it is shown in Figure 5.5.

The studied material characteristics were allocated to the model, according to Figure 5.6.

Table 5.5 Experimental data on roughing EDM (*R*) and finishing (*F*)

Probe no., Material type	EDM working modes	Machining time, t_m (min)	Machining depth, t (mm)	Removed volume, V (mm^3)	Machining rate, V_w (mm^3/min)	Surface roughness, R_a (μm)
P1, NE	*R–EDM*	6.67	1.45	36.25	5.43	1.285
P2, SOFT	*R–EDM*	9.80	1.45	36.25	3.70	1.233
P3, LUKA	*R–EDM*	7.12	1.45	36.25	5.09	1.214
P1, TiAl	*R*–EDM	12.90	0.10	2.5	0.19	1.451
P2, TiAl	*R*–EDM	8.58	0.15	3.75	0.44	1.376
P3, TiAl	*R*–EDM	11.50	0.15	3.75	0.33	1.522
P4, TiAl	*R*–EDM	10.53	0.18	4.5	0.43	1.395
P5, TiAl	*R*–EDM	7.74	0.18	4.5	0.58	1.408
P1, NE	*F–EDM*	20.97	0.05	1.25	0.06	0.483
P2, SOFT	*F–EDM*	25.52	0.05	1.25	0.05	0.435
P3, LUKA	*F–EDM*	15.44	0.05	1.25	0.08	0.390
P1 TiAl	*F*–EDM	36.44	0.05	1.25	0.03	0.801
P2 TiAl	*F*–EDM	23.20	0.05	1.25	0.05	0.693
P3 TiAl	*F*–EDM	17.05	0.05	1.25	0.07	0.898
P4 TiAl	*F*–EDM	36.64	0.05	1.25	0.03	0.742
P5 TiAl	*F*–EDM	13.17	0.05	1.25	0.09	0.773

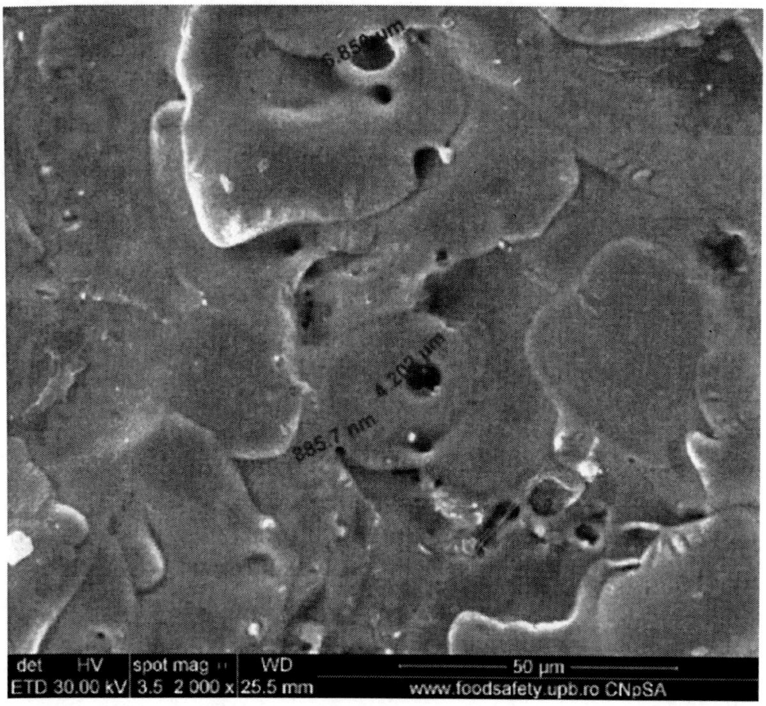

Figure 5.2 CoCr-based alloy machined by EDM finishing

Figure 5.3 Ti aluminides machined by EDM finishing

▾ **Parameters**

Name	Expression	Value	Description
hp	3[mm]	0.003 m	sample height
lp	7[mm]	0.007 m	sample width
acr	6.5e-6	6.5E-6	crater initial radius
Ra	1.2e-6	1.2E-6	Ra roughness
rbg	0.1[mm]	1.0E-4 m	gas bubble radius
tif	0.8e-6	8.0E-7	pulse time final
TfTiAl	3287	3287	Ti aluminides boiling temperature
TfCoCr	3000	3000	CoCr alloy boiling temperature
rms	0.25e-6	2.5E-7	resolidified material radius
rcp	0.0014*ti^0.5	0	channel plasma radius, time depedent
I	0.2[A]	0.2 A	discharge current
U	20[V]	20 V	discharge voltage
P	U*I	4 W	discharge power
HF	3/(3.14*1.25e-6^2)	6.115E11	thermal flux
a	0.5e-6	5.0E-7	plasma channel initial width
ti	0	0	pulse time sweep

Figure 5.4 Parameters of the EDM removal mechanism

The boundary conditions were set as heat flux inside the time-dependent plasma channel (Figure 5.7), considering its major part allocated to the workpiece since the negative polarity was chosen (Ghiculescu, 2013).

Thermal insulation is produced by gas bubbles around the plasma channel due to the vaporization of dielectric liquid and materials couple (tool and workpiece) involved in the EDM process (Figure 5.8).

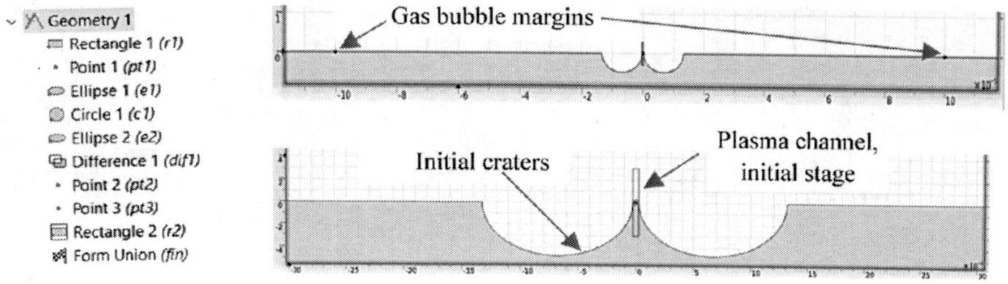

Figure 5.5 Model geometry in the zone of single discharge simulation. a) CoCr-based alloy thermo-physical properties; b) Ti aluminides thermo-physical properties

▾ Material Contents

	Property	Name	Value	Unit
✓	Density	rho	1000	kg/m^3
✓	Thermal conductivity	k	9.4	W/(m*K)
✓	Heat capacity at constant pressure	Cp	390	J/(kg*K)

(a) Co Cr based alloy thermo-physical properties

▾ Material Contents

	Property	Name	Value
✓	Heat capacity at constant pressure	Cp	520[J/(kg*K)]
✓	Density	rho	4000[kg/m^3]
✓	Thermal conductivity	k	55[W/(m*K)]

(b) Ti aluminides thermo-physical properties

Figure 5.6 Material characteristics assignment

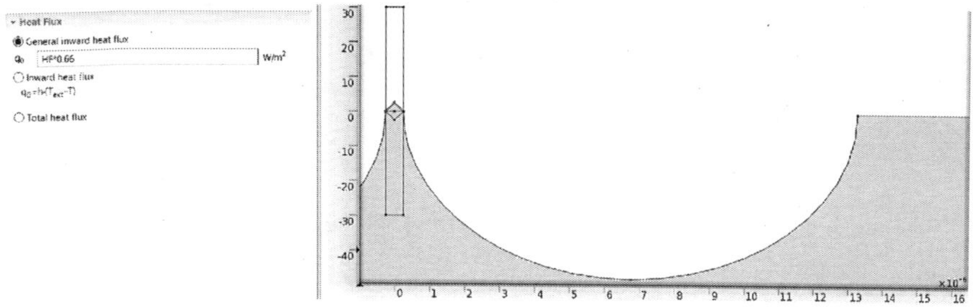

Figure 5.7 Material heat flux within plasma channel time dependent

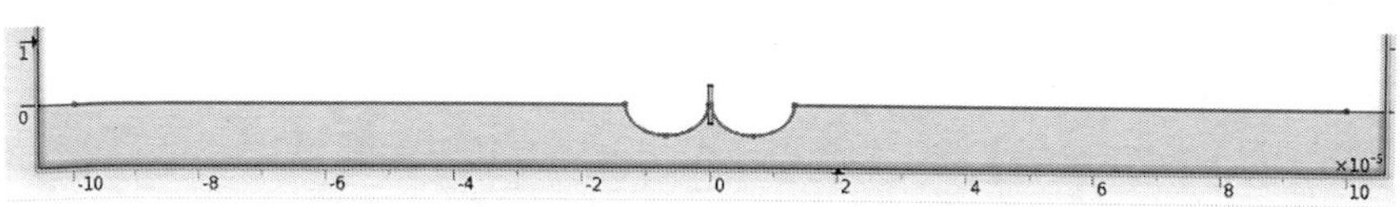

Figure 5.8 Thermal insulation due to the gas bubble

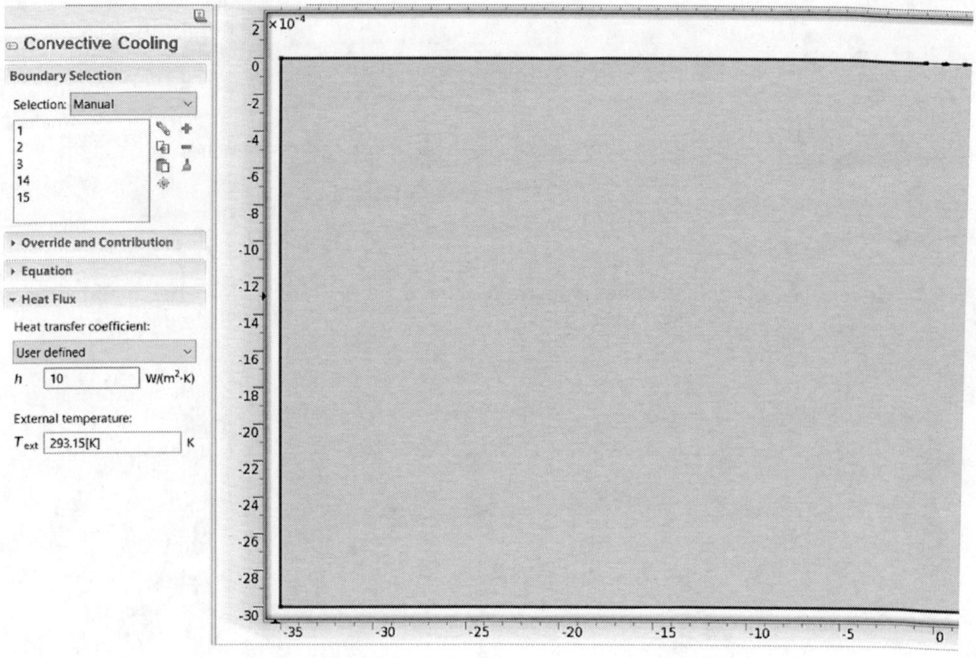

Figure 5.9 Convective cooling due to dielectric liquid

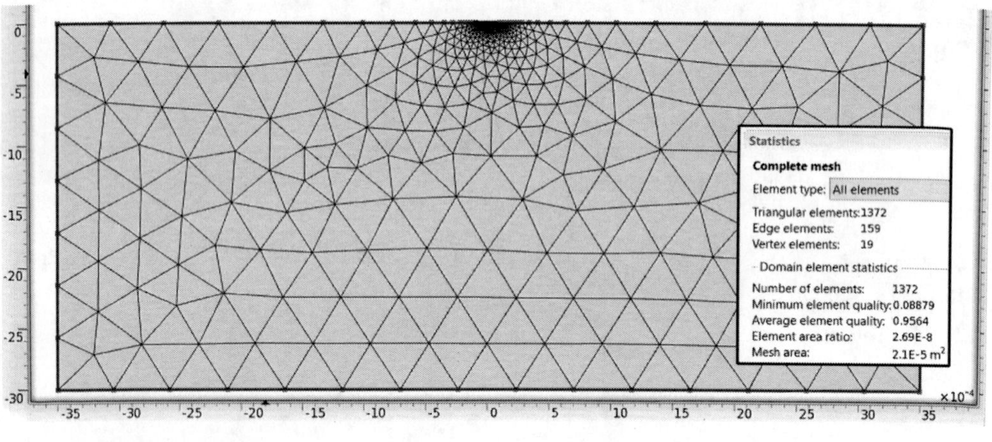

Figure 5.10 The mesh characteristics a) Boiling isothermal position in CoCr alloy sample-radius and depth; b) Boiling isothermal position in Ti aluminides sample-radius and depth

Convective cooling due to dielectric liquid surrounding the machined workpiece is illustrated in Figure 5.9.

The mesh with free triangular elements and its characteristics are presented in Figure 5.10.

The finite element model was solved using a parametric sweep variable as pulse time, *ti*, and a time-dependent one as pulse time final as *tif.*

The results in terms of temperature distribution are presented in Figure 5.11. The values exceeding boiling point indicated that this zone of material was removed.

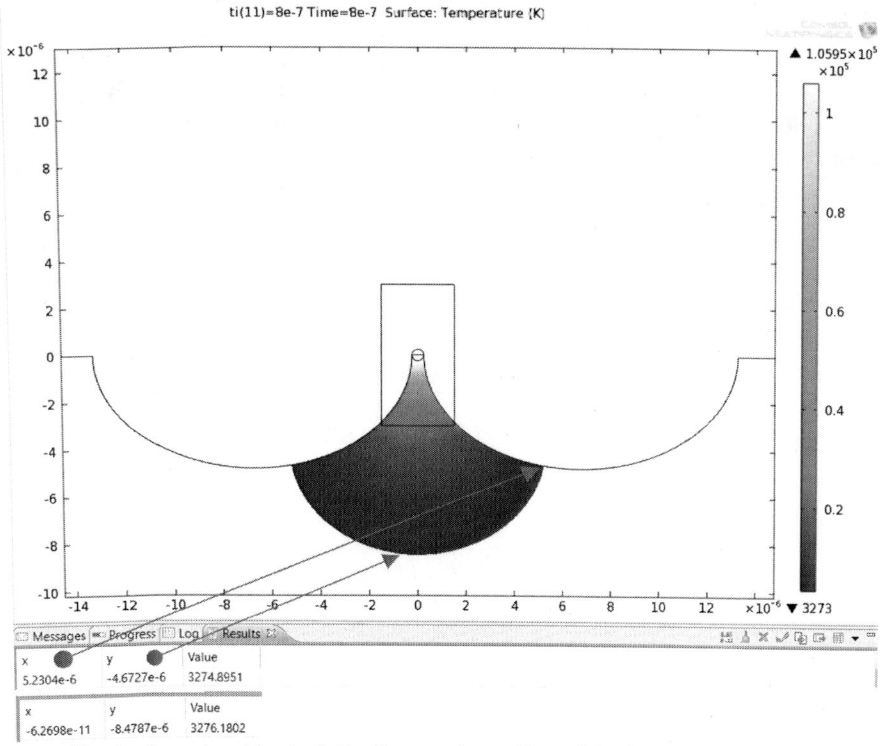

(a) Boiling isothermal position in CoCr alloy sample – radius and depth

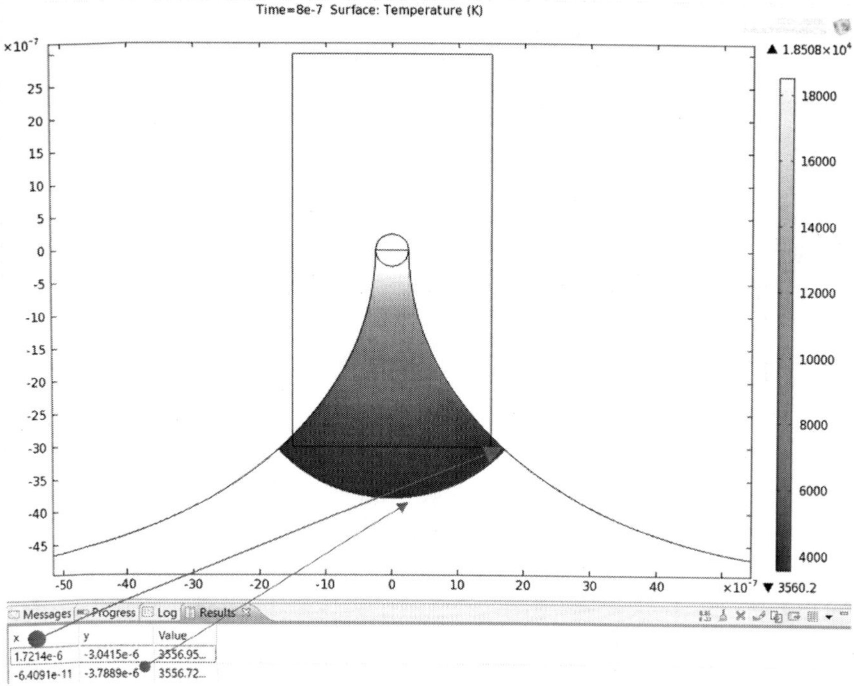

(b) Boiling isothermal position in Ti aluminides sample – radius and depth

Figure 5.11 Thermal distribution after a relaxation pulse, single discharge

Discussion

According to the overheating model, the main mechanism of material removal at EDM remains boiling. This is because the gas bubble surrounding the plasma channel lasts much longer after the pulse end. This represents a barrier against the hydraulic forces of dielectric liquid which cannot remove the melted material during discharge and after its end. So, due to overheated material above boiling temperature, that is, around 200–300°C due to high pressure generated by the plasma channel, the removal by boiling is produced since the pressure suddenly falls after the pulse end (Ghiculescu, 2013). As can be observed, the material removed by the single discharge produced by the relaxation pulse at finishing mode is much greater in the case of the CoCr alloy sample than in the case of Ti aluminides. This is in agreement with the experimental results from Figure 5.7, validating the model.

Modeling and Simulation of Deformations in the Feed System for Ultrasonically Aided Micro-EDM

The spark gap (the space between the tool and the workpiece) is dependent on voltage, amperage, material of the electrodes, and dielectric type. EDM uses high voltage and amperage for gross material removal. The higher the values for voltage and amperage the bigger the working gap becomes.

For a stable EDM process, the spark gap needs to be regularly cleaned of debris by using various flushing techniques (Makenzi et al., 2012). Failure to do so will result in debris accumulation, which leads to a short circuit state of the process, which affects its stability and consequently the surface of the tool and workpiece.

Micro electrical discharge machining (also known as µEDM) is a scaled-down EDM process. Therefore the electrode tool is much smaller and consequently more fragile than its EDM counterpart. This reduces the types of usable flushing techniques and also limits the rest. It also limits the amperage and voltage, in order to prevent excessive wear of the electrode tool (Liu et al., 2016). Small values for amperage and voltage result in a small quantity of removed material, hence low productivity, but result in a higher quality surface (Paulo, 2013).

µEDM is in an industry attention spot because of its ability to achieve surface finishing up to micro-levels (Mehdi et al., 2017). Therefore, researchers are seeking solutions to reduce its shortcomings, mainly low productivity (because of the small work gap and frequent short circuits).

Ultrasonically aided micro electrical discharge machining (µEDM+US) is one of the solutions researchers turned to. Aiding the electrode with ultrasonic vibrations in order to solve the problem of work gap flushing. When the electrode moves up and down, it creates a pump action in the work gap, effectively forcing the removed material out (Ghiculescu et al., 2009).

Being a process used for micro-levels of precision, a fast response and high-precision feed system is required to position the electrode.

One such system is presented in Figure 5.12, capable of executing a linear movement in increments of 0.2 µm.

This chapter does a numerical simulation of the roller screw's deformations in order to see its impact on the precision and on the ability of the feed system to maintain a stable µEDM+US process.

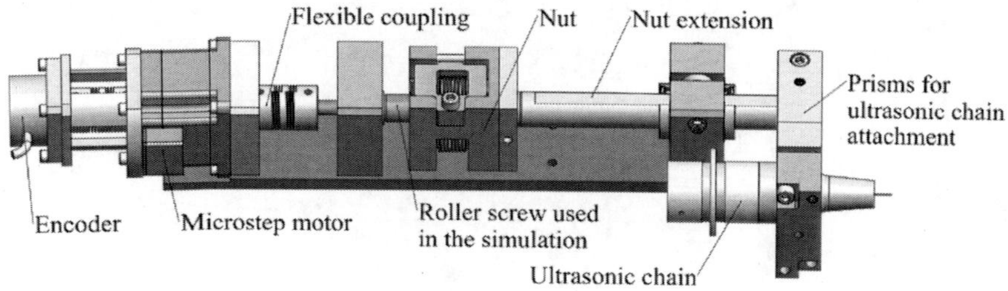

Figure 5.12 Feed system for ultrasonic aided micro electrical discharge

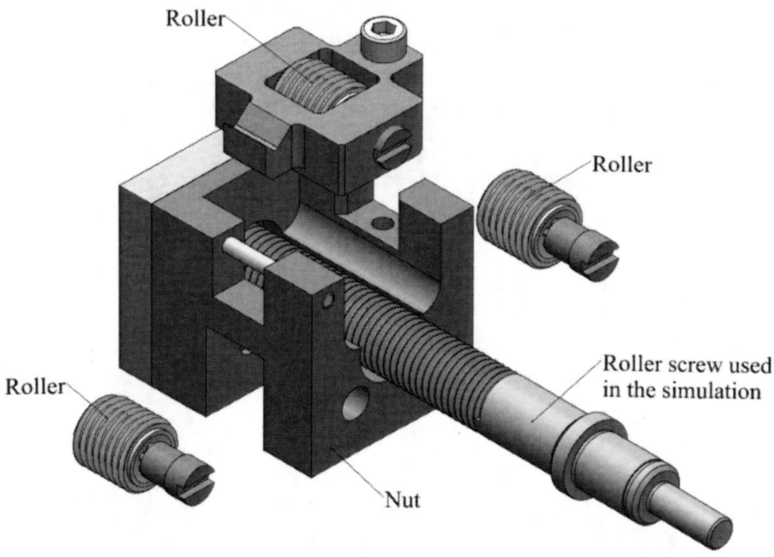

Figure 5.13 Roller screw and nut assembly

The feed system has the following characteristics: overall dimension of 478 × 129 × 66 mm, a 70 mm stroke, an angular increment of stepping motor of 0.36° degrees, a torque of 0.8 Nm, a linear increment of 200 nm (0.02 mm), and a mass of approximately 10 kg.

Contact Surface

To evaluate the deformation of the roller screw, the contact surfaces need to be identified and isolated. From Figure 5.13, it can be seen that the nut houses three rollers of size M16x2, which are the only parts that maintain contact with the roller screw.

The placement of the rollers is shown in Figure 5.14, where two are placed symmetrically at the bottom (at an angle of 45°), and one at the top, placed at an angle of 7°.

The actual contact surface is on a small zone on the thread of the screw, determined by using the command "interference detection" in Solidworks™ 2016.

The results are shown in Figure 5.15, where eight contact surfaces of approximately 0.02 mm^3 each are identified (for one roller). The measurement unit is mm^3, because the command calculates the common volume of two solid parts (roller screw and roller). The study considers this volume as the actual contact surface, approximately 0.02 mm^2 times seven, for each roller.

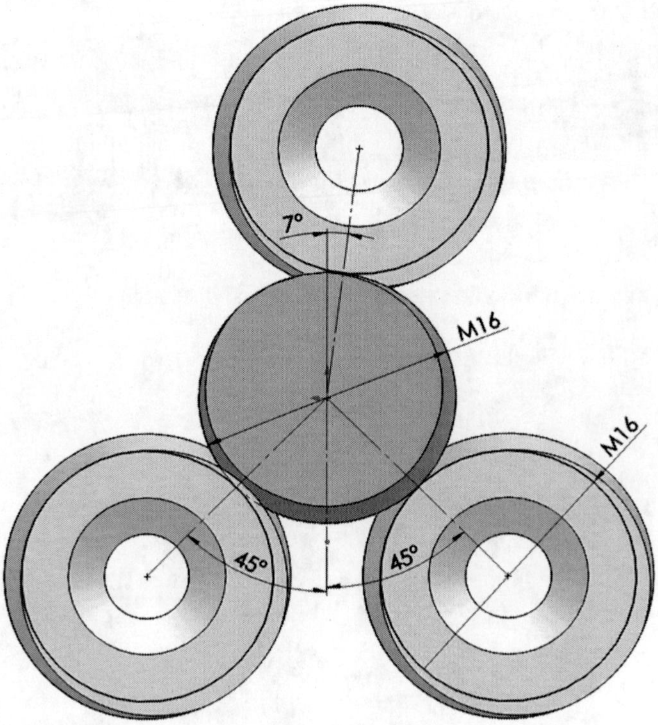

Figure 5.14 Position of the rollers around the roller screw

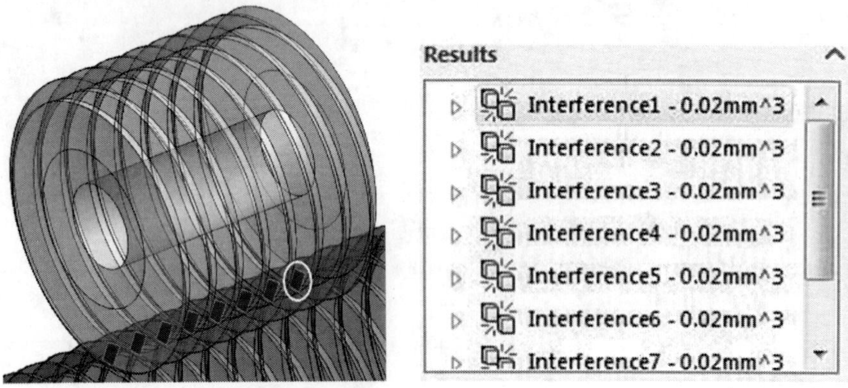

Figure 5.15 Interference detection and representation

The last thread (Figure 5.15, highlighted in white) is a variable contact surface. Meaning it's not always in contact and doesn't always have the same contact surface as the other 7. Therefore, it is excluded from the simulation.

In order to simulate the deformation of the roller screw, the contact surfaces, which have been identified, need to be isolated. Using Solidworks™ 2016, three longitudinal planes are created, at the same angles as those in Figure 5.14, in order to determine the actual contact surface and isolate it.

On every plane created, the sketch in Figure 5.17 is made, using the command "convert entities" on the flanks of the screw. This sketch is created at a distance of 35.5 mm (Figure 5.16), which is the start of the usable contact area of the nut (the stroke).

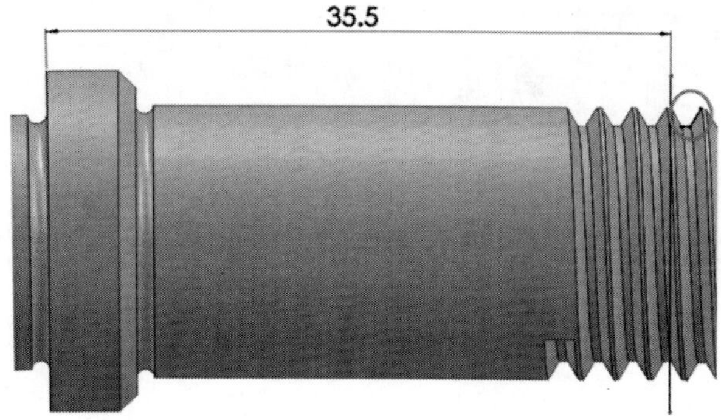

Figure 5.16 Sketch location

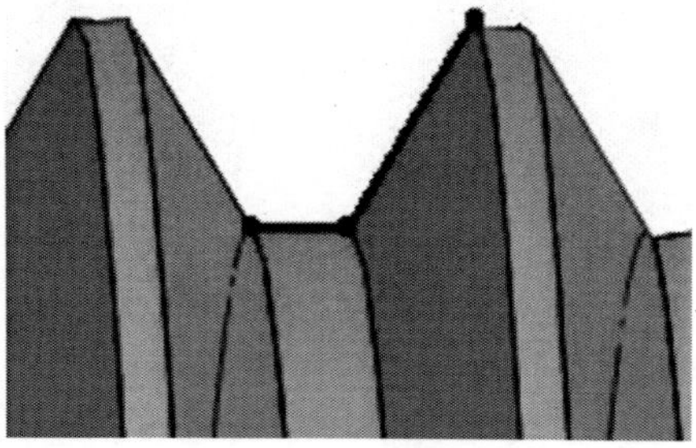

Figure 5.17 Sketch construction

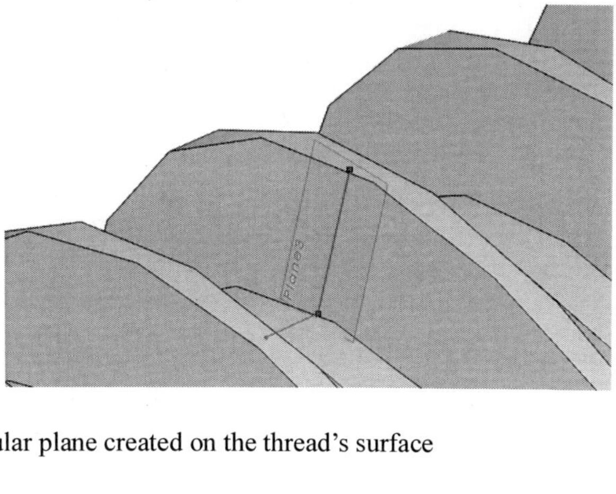

Figure 5.18 Perpendicular plane created on the thread's surface

In the sketch from Figure 5.17, a plane perpendicular to the flank of the thread is created, through the highlighted green line in Figure 5.18.

On the newly created plane, the sketch in Figure 5.19 is drawn.

These dimensions were determined by constantly comparing the isolated surface with the surface indicated by the command "interference detection." Through iterative modifications of the dimensions presented in Figure 5.19, the difference is narrowed down to 0.00164 mm^2, as seen in Figure 5.20.

Once the surface for one roller is isolated (Figure 5.20, highlighted), a circle is drawn on the roller screw that goes through the middle of the highlighted center of the rectangle previously created in Figure 5.19.

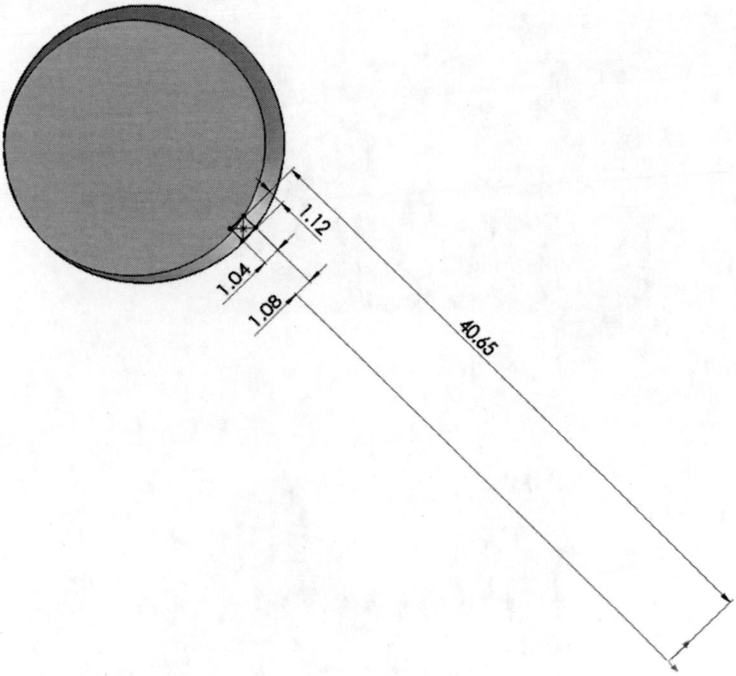

Figure 5.19 Sketch dimensions for surface location

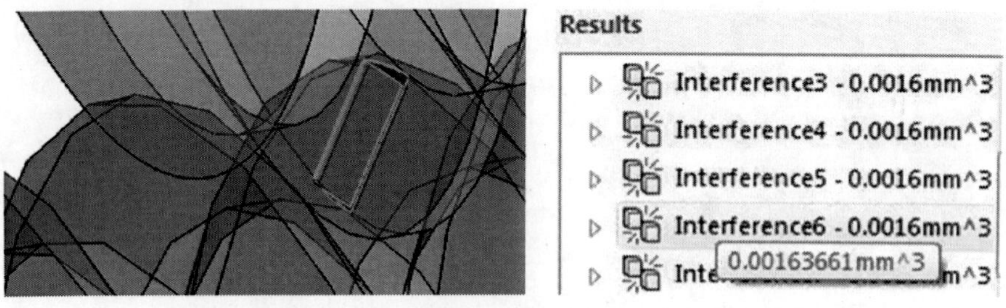

Figure 5.20 Isolated surface compared to interference detection

Using this circle and the dimensions from Figure 5.19 (1.12 and 1.04 mm), the dimensions from the origin (screw center) from Figure 5.19 (40.65 mm and 1.08 mm) are modified such that the middle points of the other two rectangles are situated on the previously drawn circle, as seen in Figure 5.21. The isolated surface is always compared to the surface indicated by the command interference detection and narrowed down as much as possible.

After isolating the first contact surface for each roller, the command "LPattern" is used to copy them onto the next six flanks, at a distance of 2 mm (the pitch of the roller screw), as seen in Figure 5.22.

Since the rollers don't have a stationary position, and are moving along the roller (within the stroke area), two more sets of contact surfaces are added, as shown in Figure 5.23. The same command "LPattern" is used, and the isolated surfaces previously built are selected, to create position 2 and position 3. The distances at which they are created are presented in Figure 5.23.

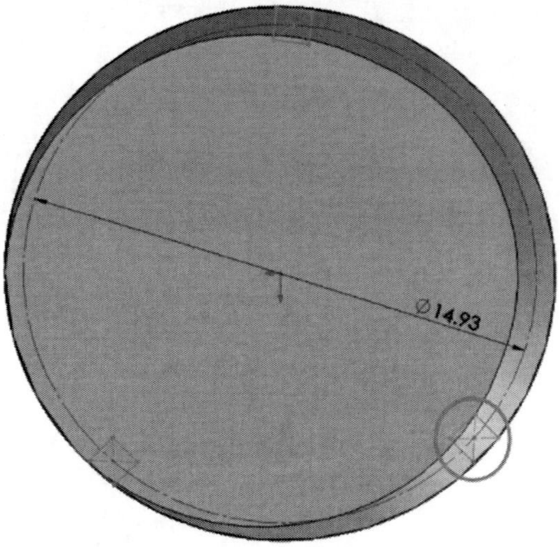

Figure 5.21 Sketch distribution of the isolated surfaces

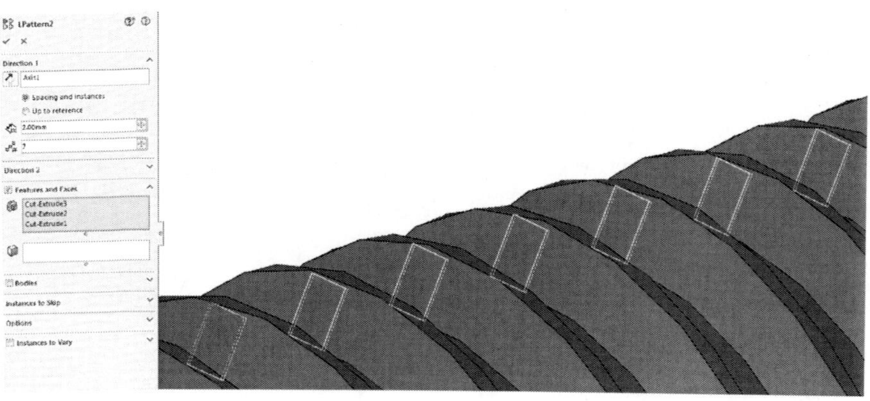

Figure 5.22 Multiplying the isolated surfaces

Simulation

The roller screw, with the isolated surfaces, is exported from Solidworks 2016 into Inventor™ 2020, where the analysis is performed (Figure 5.24).

The material of the roller screw is added to Inventor's material library (Figure 5.25), 50CrMo4—recommended by SKF™ (roller screws), with the characteristics presented in Table 5.6.

Mesh settings are set to default (Figure 5.26). The meshing process consists of splitting the geometry of the roller screw into a finite number of elements. So the solution is easier to find in small simple elements, rather than in big complex ones. The higher the meshing quality, the more accurate the simulation is.

After the meshing and the material allocation, a fixed constraint is put in place on the highlighted surface—where a bearing is assembled, as seen in Figure 5.27.

The force is added to every isolated surface in the first position (Figure 5.28).

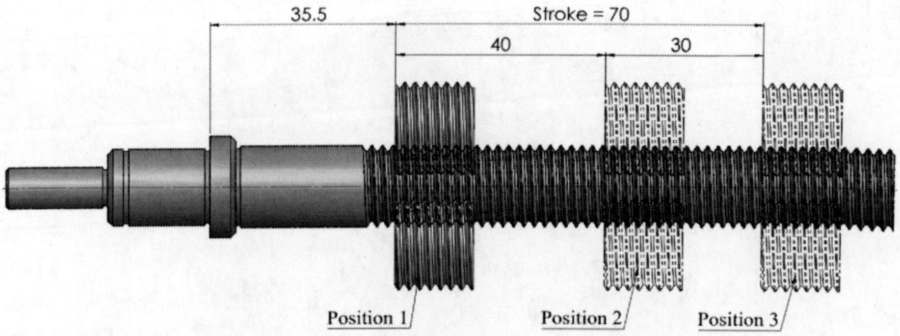

Figure 5.23 Multiple positions for testing purposes

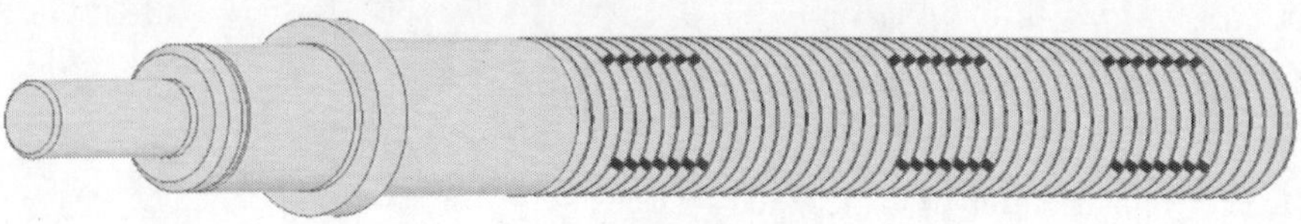

Figure 5.24 Roller screw imported in Inventor™ with highlighted contact surfaces

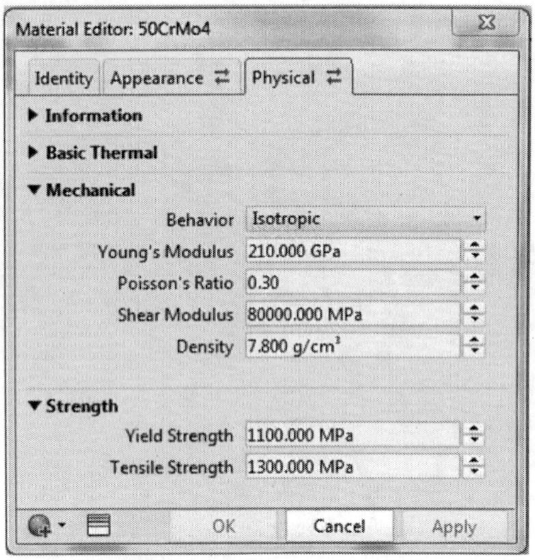

Figure 5.25 Adding 50CrMo4 to the material library and assigning it to the roller screw

The nut (and all parts attached to it) has an estimated weight of 1 Kg. Multiplying the weight with the value of gravitational acceleration (9.81 m/s^2) results in a force of 9.81 N, which is then approximated to a fair value of F = 10 N and used in the simulation. The direction of the force is along the X-axis, as shown by the arrows in Figure 5.28.

Results and Discussion

The displacement along the X-axis is shown in Figure 5.29. The maximum displacement value is 0.02363 µm. However, the three rollers on the screw are not stationary, so for the study to be relevant, the other two positions (see Figure 5.23) are simulated.

Table 5.6 Material characteristics for 50CrMo4

Code	50CrMo4
Young's modulus (GPa)	210
Poisson's ratio (–)	0.3
Shear modulus (MPa)	80,000
Density (g/cm^3)	7,800
Yield strength (MPa)	1,100
Tensile strength (MPa)	1,300

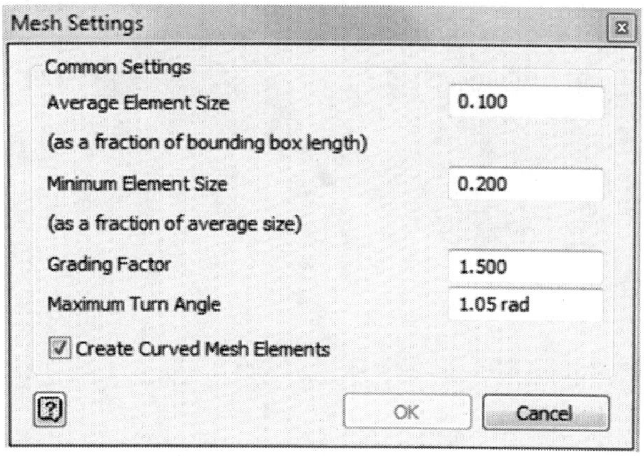

Figure 5.26 Mesh settings

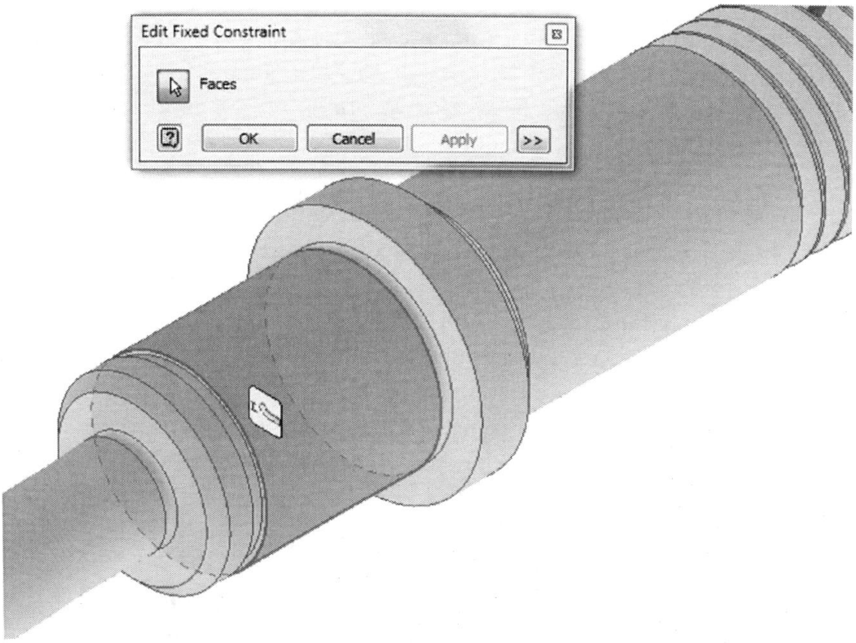

Figure 5.27 Adding a fix constraint to the highlighted surface

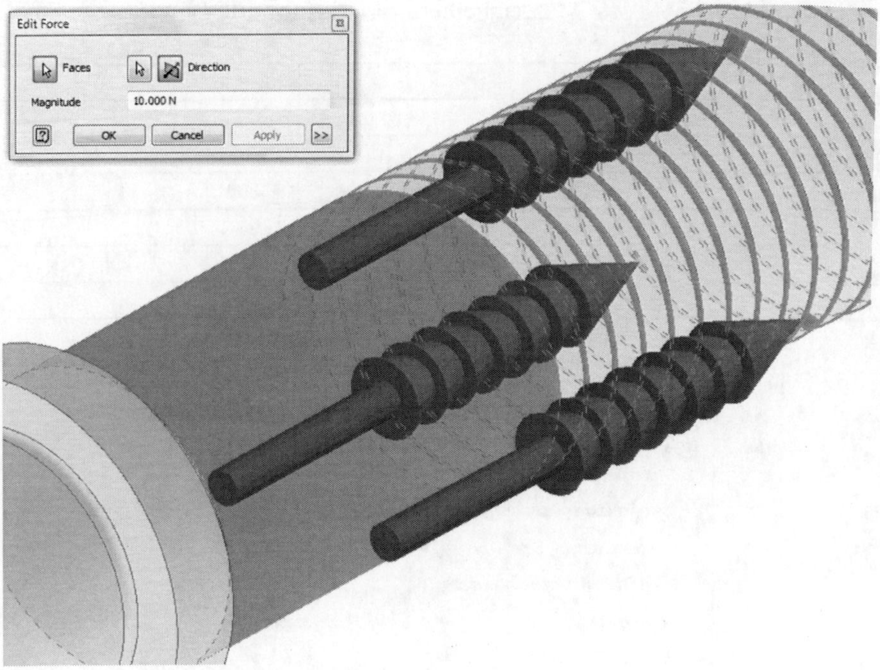

Figure 5.28 Adding a force of 10 N to the isolated surfaces

The results are presented in Table 5.7 and Figure 5.29.

The maximum deformation is found in position 3, at the end of the stroke, with a value of 0.057 μm.

In μEDM+US, the usual working gap is around 5 μm, with an amplitude of 1–2 μm modifying it. The minimum frontal gap is considered to be 1 μm, taking account of discharge parameters and particle filtration of less than 1 μm of dielectric liquid.

The feed system's accuracy is 0.2 μm. However, since, during the μEDM+US process, the position of the tool-electrode varies between two points of elongation (amplitude of around 2 μm), due to ultrasonic vibrations, the deformation of the screw will have a minimal impact on accuracy.

For comparison purposes, a second roller screw is simulated, but with the rollers being set at equidistant angles of 120° on the circumference of the roller screw.

For the second roller screw (equidistant) the deformation values are presented in Table 5.8 and its displacement along the *X*-axis in Figure 5.30.

In Figure 5.29, it can be observed that the deformation isn't evenly distributed on the roller screw. This is because the rollers are not disposed at equidistant angles, so the deformation is skewed to the right, where the 7° roller is located, hence the brighter color.

In comparison, for the equidistant rollers placement (Figure 5.30), the deformation is evenly distributed, the green color is more prominent. The deformation values are also lower for the equidistant roller, 0.018 μm vs 0.023 μm, which makes this variant ideal for mitigating the effects of deformation upon the feed system's precision.

However, the reason why the equidistant rollers are not used is that the roller placed at 7° acts as a lever and brings all rollers into contact with the roller screw with a greater closing force.

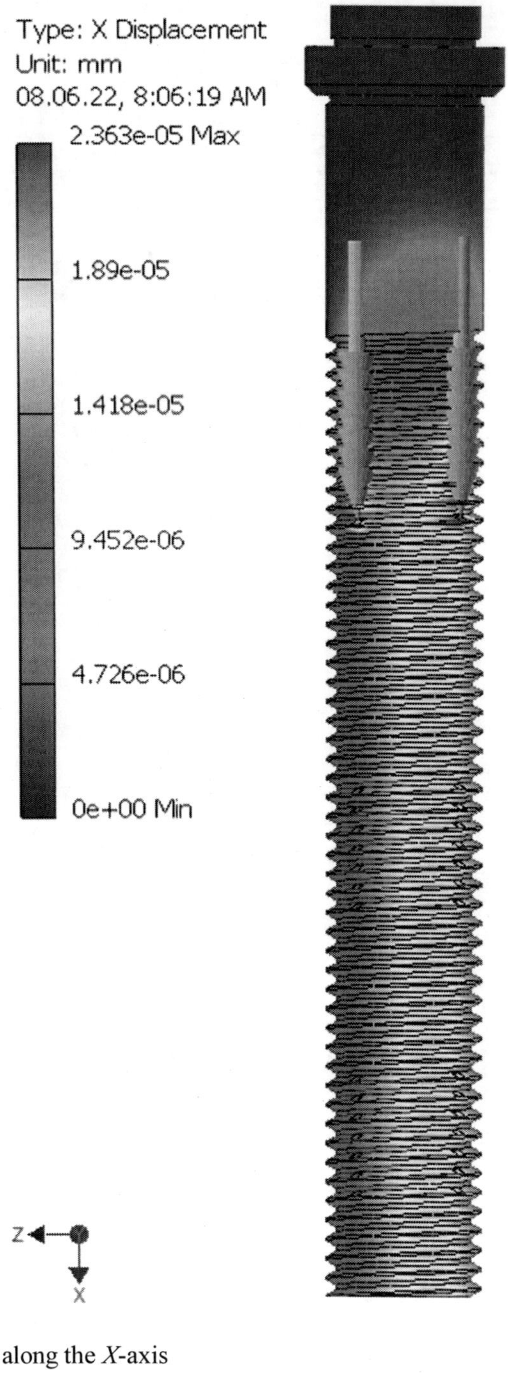

Figure 5.29 Displacement along the *X*-axis

Table 5.7 Maximum deformation along the *X*-axis

Position	Maximum value (mm)	Maximum value (μm)
1	2.363e−05	0.02363
2	4.221e−05	0.04221
3	5.727e−05	0.05727

Table 5.8 Maximum deformation along the *X*-axis for equidistant rollers

Position	Maximum value (mm)	Maximum value (μm)
1	1.848e−05	0.01848
2	3.116e−05	0.03116
3	4.08e−05	0.0408

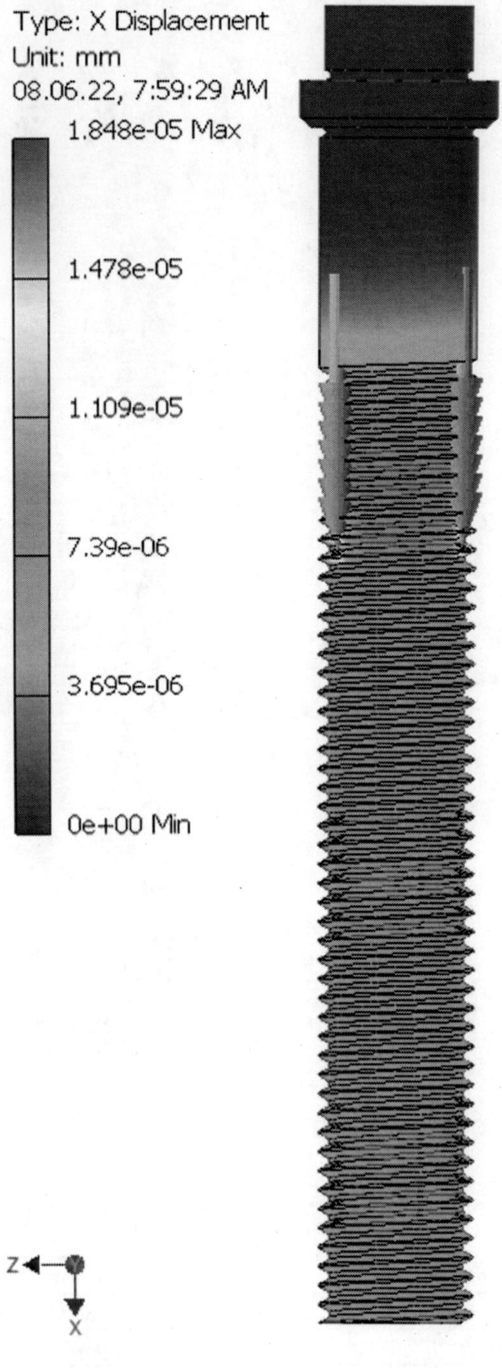

Figure 5.30 Displacement along the *X*-axis for equidistant placement of the rollers

Conclusion

For both constructive variants of the roller screw, the greatest deformation was found in position 3, at the end of the longitudinal stroke, measuring 0.057 μm for the asymmetric disposal, and 0.041 μm for the equidistant disposal, on screw circumference.

The initial screw with a 7° asymmetric roller disposal is preferred, because of its possibility to be used as a lever to bring all other rollers in close contact with a greater closing force on the roller screw. Even though the second variant has lower deformations and a more uniform variation along the roller screw.

The screw deformations, relative to the linear incremental movements of 0.2 μm of the feed system, and within around 5 μm frontal working gap, that varies with about 2 μm, due to US oscillation amplitude, are low and should not have a significant influence on the precision and stability of the process in any of the tested positions.

Further research will be done on the physical feed system, in a variety of different positions, to validate the results of the finite element analysis, using appropriate measuring instruments like laser interferometry.

Eigenfrequency Simulation for a Roller Screw Used as a Feed System for Ultrasonically Aided Micro-electrical Discharge Machining

Eigenfrequencies are certain discrete frequencies at which a part or system is prone to vibrate. If a system is actuated by an exterior periodical force, whose frequency is equal or close to one of the natural frequencies of the system or parts of it, the produced vibration has relatively big amplitudes, even for relatively small amplitudes of the perturbing force. This represents the resonance state of the system (Rades, 2008).

This is a case to be studied for the feed system for ultrasonically aided micro-electrical discharge machining—μEDM+US, which requires good control of the process in terms of positioning precision and the time required to achieve the commanded position. Considering that the μEDM+US process takes place within a very narrow work gap of less than 5 μm, between the surface of the tool (which vibrates with ultrasonic frequency with an amplitude of around 1–2 μm) and surface of the workpiece, the problem of the resonance of the feed system becomes critical (Jahan et al., 2015).

The study is made to see if the perturbing force, induced by the stepping motor that actuates the feed system, could produce resonance by having its command frequency overlap on the own frequency of some parts of the system, especially the roller screw, the most important part of the system (concerning the effective linear increment).

The natural frequency of the roller screw is studied, using a dedicated module for eigenfrequency calculation, considering chiefly the torsional vibration that could be induced by the stepping motor.

One study (Diego et al., 2012) addresses the modeling and vibration mode analysis of a ball screw drive using a dynamic model to examine the resonant modes of the system that could be excited, which led to degrading the system's accuracy.

In roller screw systems, numerous elements contribute to the flexibility of the system, and this flexibility contributes to vibratory behavior in any mechanical system (Orang et al., 2009).

Torsion vibrations were studied by Zhou et al. (2007) using the finite element method (FEM). It resulted in the discovery that the table position, relative to the roller screw, changes all the natural frequencies of the roller screw, similar to the results presented in the current study.

Roller Screw Design

The study is made to see if the roller screw assembly enters in resonance with the frequency used by the command system of the motor, within a feed system designed for μEDM+US (Figure 5.31). Since the nut is complex and has many parts, the simulation contains only the roller screw and the three rollers (named roller screw assembly) that are in contact with the roller screw for ease of calculations.

Motor Frequency

The motor used in the assembly is a high-accuracy stepper motor (Oriental Motor General Catalogue, C-121).

In Figure 5.32, the speed and torque characteristics of the motor are presented. In order to ensure maximum acceleration, the torque must have maximum values. Maximum acceleration is required in order to minimize the response time of the feed system in case of critical situations.

The maximum torque is achieved between the frequencies 0.05–30 kHz, as seen in Figure 5.33, highlighted in blue. The higher the frequency, the faster the rotor of the motor rotates.

The maximum speed can never be exceeded in any given situation. In the μEDM+US process, it's used to avoid short-circuits and idle machining time.

The minimum speed is used to provide the linear displacement for technological feed and to determine the origin of the axis and the origin of the part. Both speeds are calculated using formulas:

$$v_{max} = i * f_{max} \left(mm/s\right) \tag{3}$$

$$v_{min} = i * f_{min} \left(mm/s\right) \tag{4}$$

where *i*—linear increment of the feed system (mm); f_{max}—maximum frequency of the pulses (Hz); f_{min}—minimum frequency of the pulses (Hz).

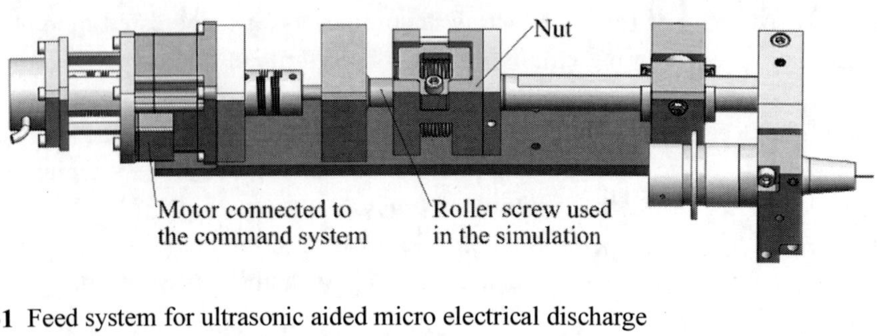

Figure 5.31 Feed system for ultrasonic aided micro electrical discharge

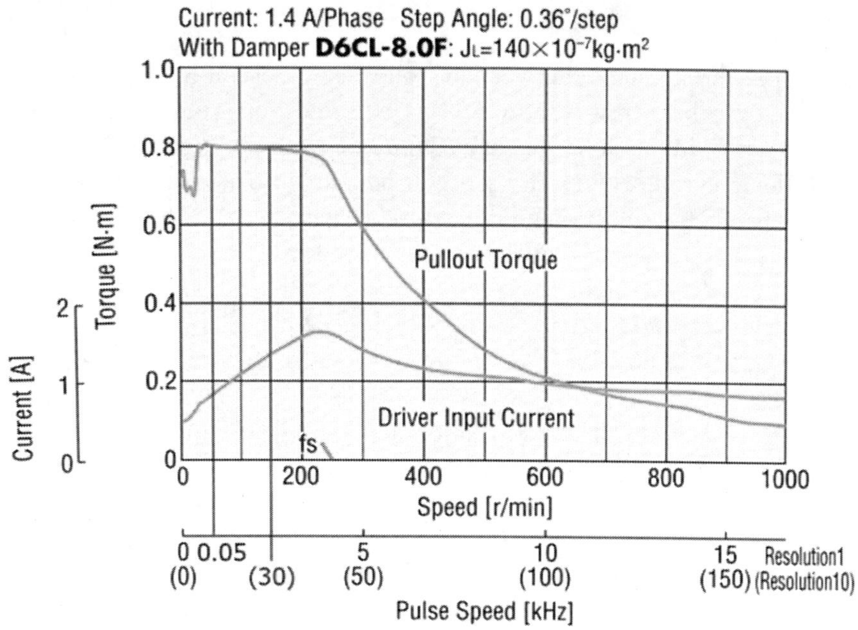

Figure 5.32 Speed–torque characteristics of the stepper motor CRK564PMBP (Oriental Motor General Catalogue)

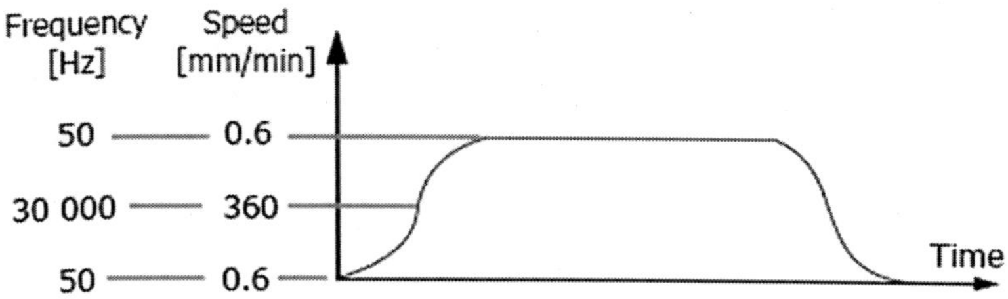

Figure 5.33 Speed profile (*S*), used by the motor

At every pulse, the feed system executes a movement of 0.2 μm, so $i = 2 \times 10^{\times 4}$ mm. The maximum frequency at which the torque is maximum is 30,000 Hz, and the minimum is 50 Hz, therefore $f_{max} = 30{,}000$ Hz and $f_{min} = 50$ Hz.

Replacing the values in formulas 1 and 2, the maximum speed, $v_{max} = 360$ mm/min, and the minimum speed, $v_{min} = 0.6$ mm/min, of the motor are obtained.

S Speed Profile

The motor, using its programmed software, accelerates and decelerates using an "S" speed profile, as seen in Figure 5.33. This way it avoids shocks caused by sudden acceleration and deceleration.

The command system mainly uses a frequency of 50 Hz while machining. The frequency of 30,000 Hz is used in case of a short-circuit, out of range for sparking and to rapidly advance toward the workpiece. From these results the 30,000 Hz and the 50 Hz frequencies are the most used.

Simulation

In order to determine the resonance between the roller screw assembly and the frequencies of the command system, the roller screw (Figure 5.34) is exported from Solidworks™ and imported in Comsol Multiphysics™, in the Solid Mechanics module, with the Eigenfrequency study.

Afterward, the material of the roller screw is added in Comsol (Figure C-5), 50CrMo4—recommended by SKF™ (A selection guide), with the characteristics presented in Table 5.9.

Afterward, the material of the roller screw is added in Comsol™'s library (Figure 5.35), 50CrMo4—recommended by SKF™.

The material characteristics are presented in Table 5.9.

The meshing is set to fine as seen in Figure 5.36, where the number of elements and their minimum quality is shown.

A visual representation of the mesh quality is displayed in Figure 5.37, Number 1, color-coded green is an element with the ideal quality, while 0, color-coded red is an element with the worst quality. The better the quality, the more accurate the simulation.

Going further on the model builder tree, the configuration of the eigenfrequency study is set (Figure 5.38).

In this case, the 50 Hz test frequency is typed in "search for eigenfrequencies around." The result is presented in Figure 5.39, where the closest value is 0.012 Hz. In this case, the roller screw does not enter resonance with the command system's frequency.

The rollers on the roller screw are not stationary. Therefore the eigenfrequency needs to be tested when the rollers are in several different positions (Figure 5.40).

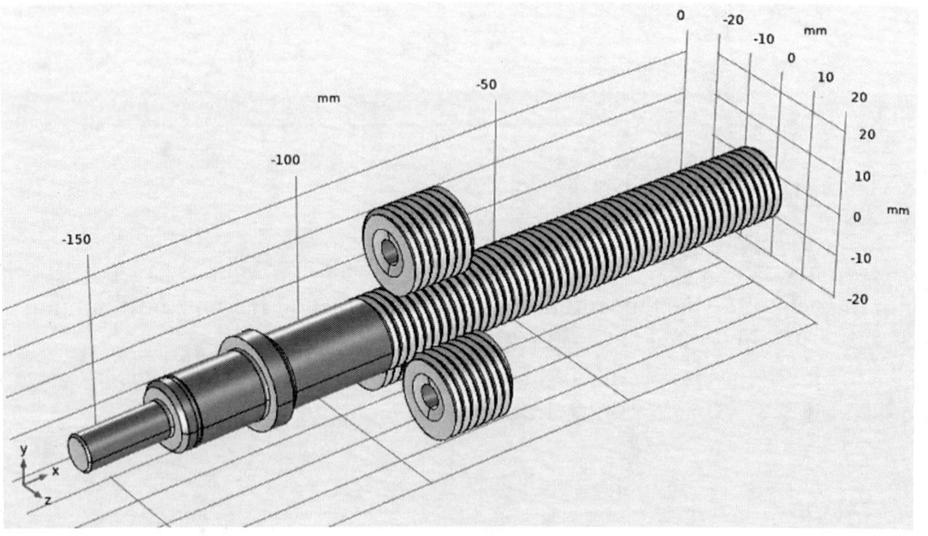

Figure 5.34 Roller screw assembly imported in Comsol Multiphysics

Table 5.9 Material characteristics for 50CrMo4

Name	Density (kg/m³)	Young's modulus (Pa)	Poisson' ratio (–)
50CrMo4	7800	210×10^9	0.3

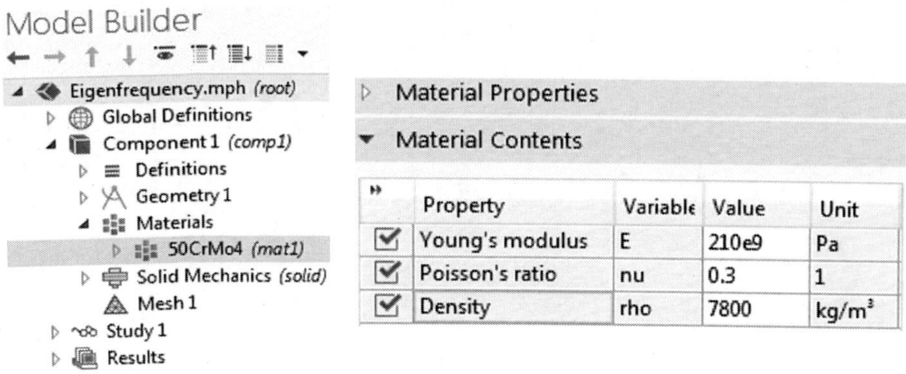

Figure 5.35 Adding 50CrMo4 as the material of the roller screw

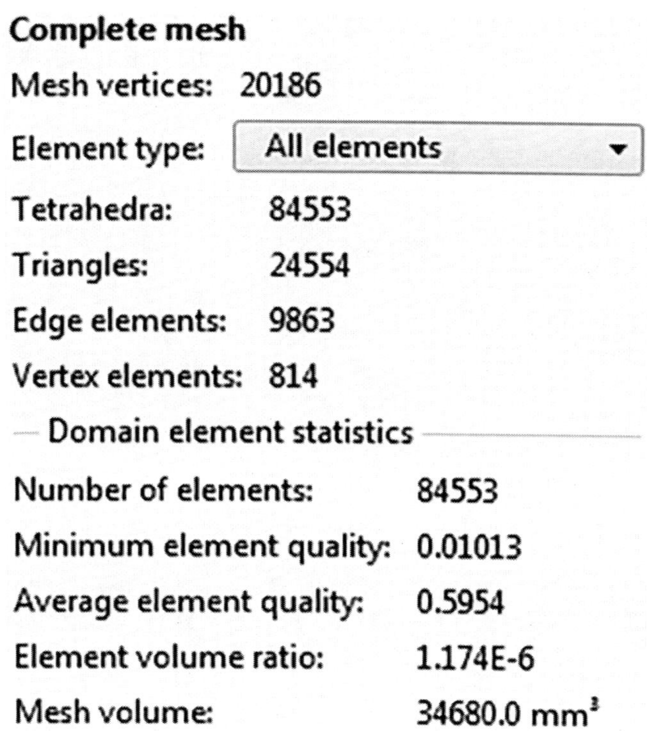

Figure 5.36 Element statistics

Results and Discussion

The set of values (eigenfrequency) has been added in Table 5.10, for all positions, in order to check if the roller screw assembly enters resonance. The testing frequencies are 50 Hz and 30,000 Hz, respectively.

The values bolded are the closest to the testing frequencies (Table 5.10). The minimum difference, for the testing frequency of 50 Hz is more than 49.9 Hz. For the testing frequency of 30,000 Hz, the minimum difference is more than 200 Hz.

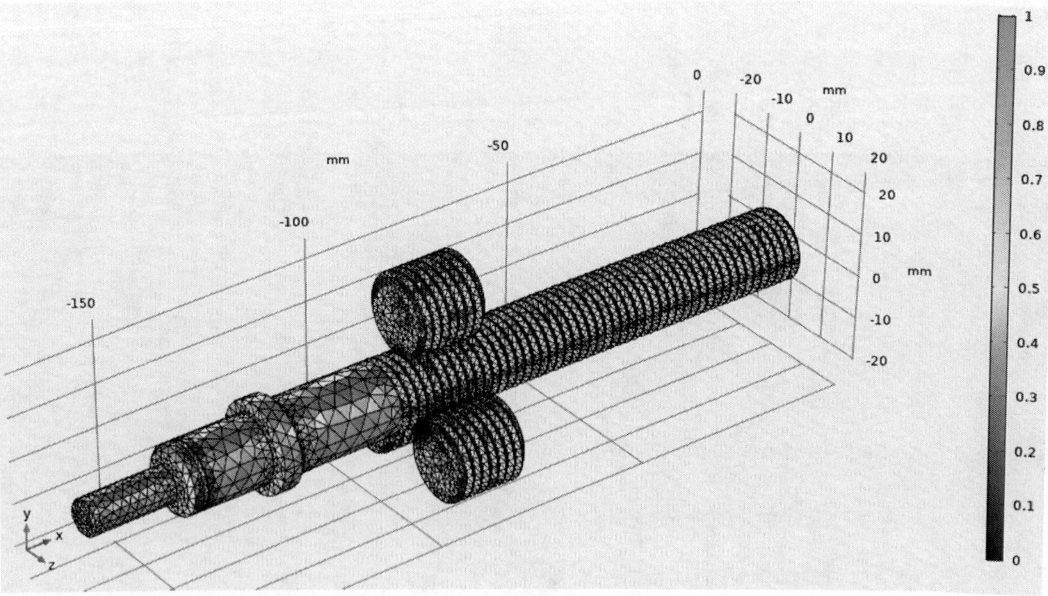

Figure 5.37 Mesh quality of the roller screw

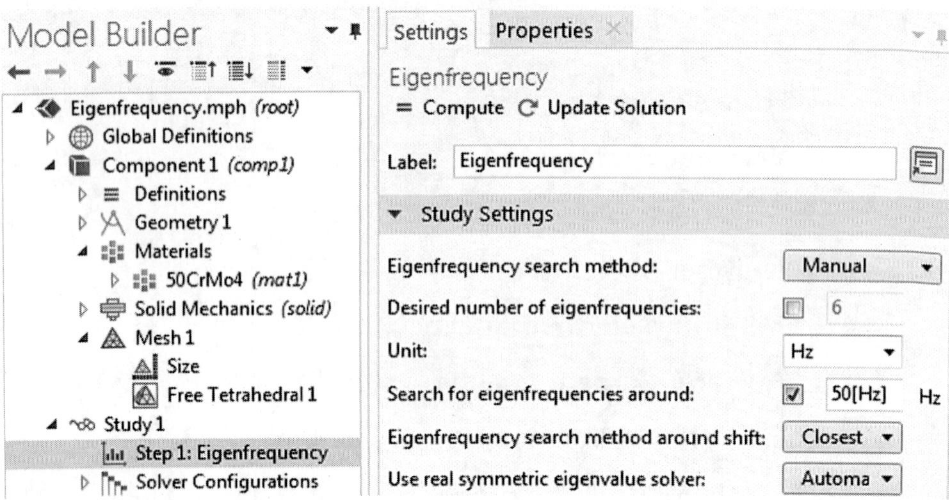

Figure 5.38 Eigenfrequency study

The position of the rollers changes the natural frequencies of the roller screw, as seen in Table 5.10. However, these positions are not static, rather transitory, it can be stated that the designed feed system does not enter resonance during machining. And if it enters, it is only for a brief period and should not affect the feed system's accuracy.

The only exception is in position V, where there's a difference of 27 Hz. This would mean that the feed system is advancing or retracting at maximum speed at the end of the stroke, which is not possible, as it uses an *S* speed profile as seen in Figure 5.33.

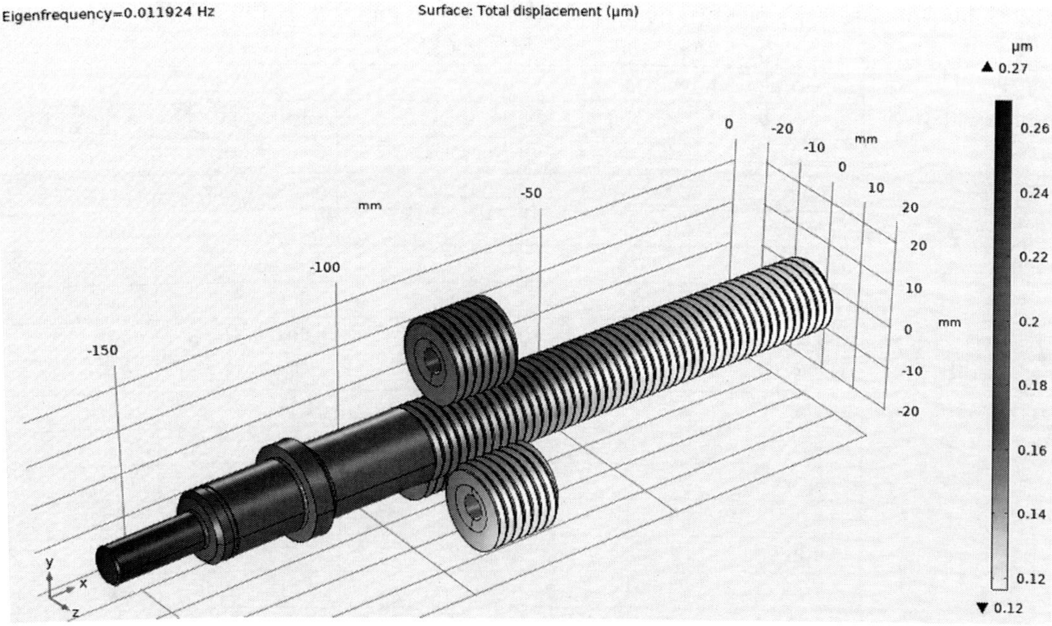

Figure 5.39 Roller screw eigenfrequency study on a frequency of 50 Hz

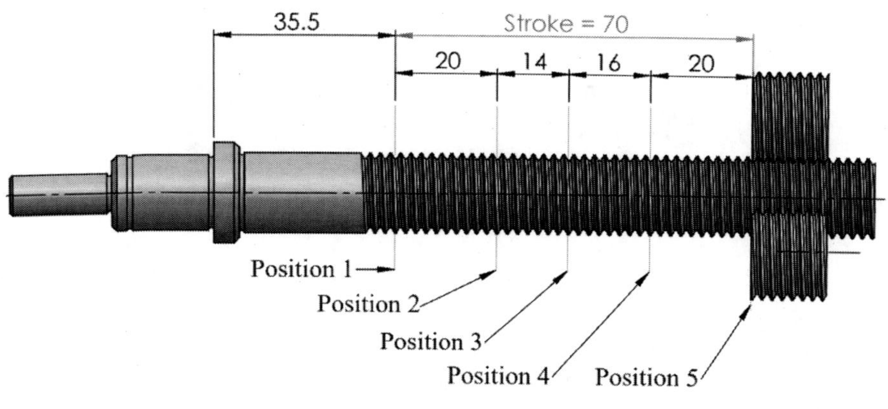

Figure 5.40 Roller screw with rollers in different positions

After dynamic modeling tests using the player command from Comsol™, the frequencies presented in Table 5.10, indicated only longitudinal vibration modes, but the stepping motor is expected to also induce torsional vibration modes.

Conclusion

Considering that the material removal process takes place in a very narrow working gap of less than 5 μm, the eigenfrequency simulation becomes important. Induced vibration (due to resonance) negatively influences machining accuracy, in an already very narrow work gap. Based on the simulation, the roller screw assembly does not enter in resonance with the command

Table 5.10 Roller screw's eigenfrequency for different testing frequencies and roller positions

Testing frequency (Hz)	**Eigenfrequency (Hz)**				
	Position I	**Position II**	**Position III**	**Position IV**	**Position V**
50	0.0031i	0.0038i	0.0012i	0.0012i	0.0037i
	0.0042i	0.0049i	0.0022i	0.0028i	0.0109i
	0.0051i	0.012i	0.0047i	0.0107i	0.0009
	0.0161i	0.0025	0.0111i	0.0018	0.0025
	0.0042	0.0072	0.0022	0.0052	0.0054
	0.012	**0.012**	**0.011**	**0.009**	**0.011**
Difference (Hz)	*49.988*	*49.988*	*49.989*	*49.991*	*49.989*
30,000	**30,405**	23,763	27,689	**31,057**	26,404
	30,685	**30,218**	29,658	31,846	26,462
	32,693	30,885	**29,682**	33,020	**30,027**
	32,851	33,068	31,609	33,163	31,127
	32,910	33,488	33,935	33,713	32,904
	33,456	33,711	34,179	33,932	34,185
Difference (Hz)	*405*	*218*	*318*	*1 057*	*27*

system, as both the testing frequency and the eigenfrequency of the assembly are not equal, or close in any position of the rollers. Further research will be done on the physical feed system for the confirmation of computerized results.

Modeling of Laser Beam Cutting of Steels and Polymeric Materials

Laser cutting process is a thermal process in which the laser beam (Laser Beam Machining (LBM)) is focused and used to melt the material in order to generate contours. A coaxial gas jet is used to remove the molten material (Ghiculescu, 2004). Both the laser and the plasma cutting processes (presented below) are automated with CAD/CAM systems, usually controlling three-axis flat beds or more advanced developed six-axis robots for complex three-dimensional machining (Kumar et al., 2017).

It is recommended that laser cutting machines be operated at the maximum power and processing speed possible aiming at the protection of the working environment. On the other hand, from the consumption resources point of view, it is necessary to optimize the cut in order to save raw material and processing time and reduce energy consumption and machine wear (Radovanovic et al., 2006). Therefore, modeling and numerical simulation of the laser process before the effective cutting process becomes of utmost importance under the conditions relating to the objectives mentioned above.

In this study, this type of machining was used for cutting some steels, e.g., usual C45 (1.0503): EN 10277-2-2008 and stainless steel 305, and well as polymeric materials like PVC and plexiglass.

The experimental data obtained at laser cutting of the materials mentioned above are presented in Table 5.11.

Table 5.11 Characteristics of the laser cutting machines Bodor i7 1kW and NOVA51

Characteristics	Bodor i7 1kW	NOVA51
Laser source power (W)	1,000	100–130
Mode of operation	Continuous/modulated	Continuous/modulated
Wavelength (nm)	1080±10	1,080±10
Laser beam quality	≤1.5mm x mrad (50μm QBH)	—
Frequency (kHz)	≤20	≤5
Diameter of optical fiber (μm)	50, doped with ytterbium	—
Laser spot diameter (mm)	Focalization depending on thickness	Focalization depending on thickness
Maximum thickness (mm)	12 (steel)	10 (polymeric materials)

Figure 5.41 Sample from 304 stainless steel with 2 mm thickness cut with laser speed, $v = 80$ mm/s, power, $P = 1{,}000$ W—general view of the cut surface

Some images of the surfaces obtained by laser cutting of 304 stainless steel sample of 2 mm thickness, taken by Scanning Electron Microscope QUANTA INSPECT F50 with field emission electron gun with a resolution of 1 nm are presented in Figures 5.41–5.44.

For comparison, some images provided by QUANTA INSPECT F50 SEM of cut surfaces from C45 steel, 2 mm thickness are presented in Figures 5.45–5.48.

It can be observed the surface quality obtained at laser cutting of stainless steel is definitely superior to that resulted from C45 steel in terms of surface integrity and roughness.

The images provided by QUANTA INSPECT F50 SEM at laser cutting of PVC samples of 2 mm thickness are presented in Figures 5.49–5.52.

For comparison, images obtained at laser cutting of Polymethyl methacrylate (PMMA) – Plexiglass samples taken by QUANTA INSPECT F50 SEM are presented in Figures 5.53–5.56.

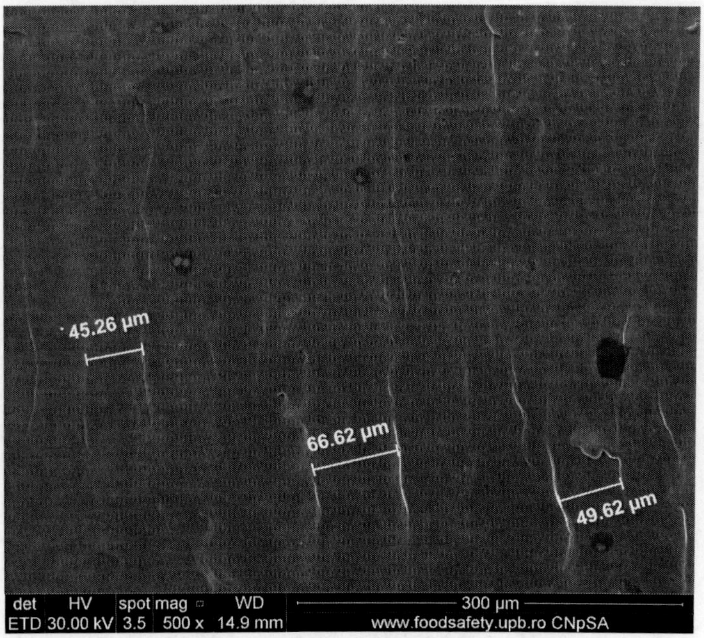

Figure 5.42 Sample from 304 stainless steel with 2 mm thickness cut with laser speed, $v = 80$ mm/s, power, $P = 1{,}000$ W—the middle of the cut surface

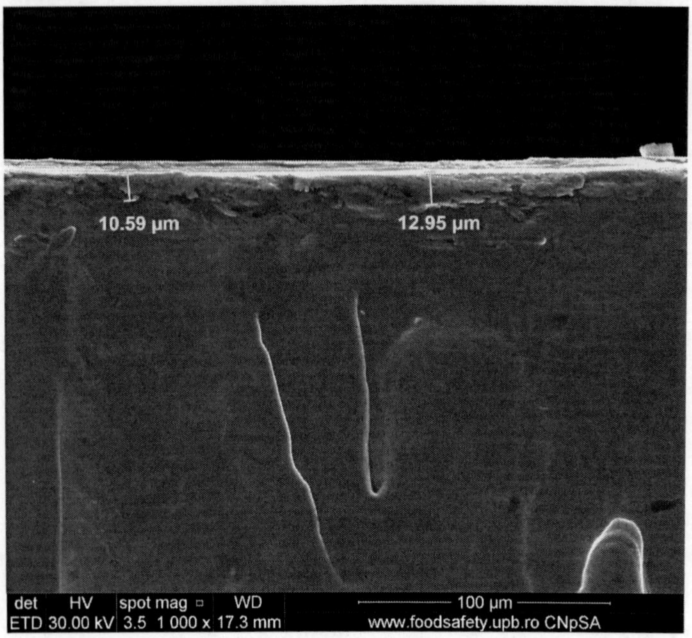

Figure 5.43 Sample from 304 stainless steel with 2 mm thickness cut with laser speed, $v = 80$ mm/s, power, $P = 1{,}000$ W—the cut surface at laser beam entrance

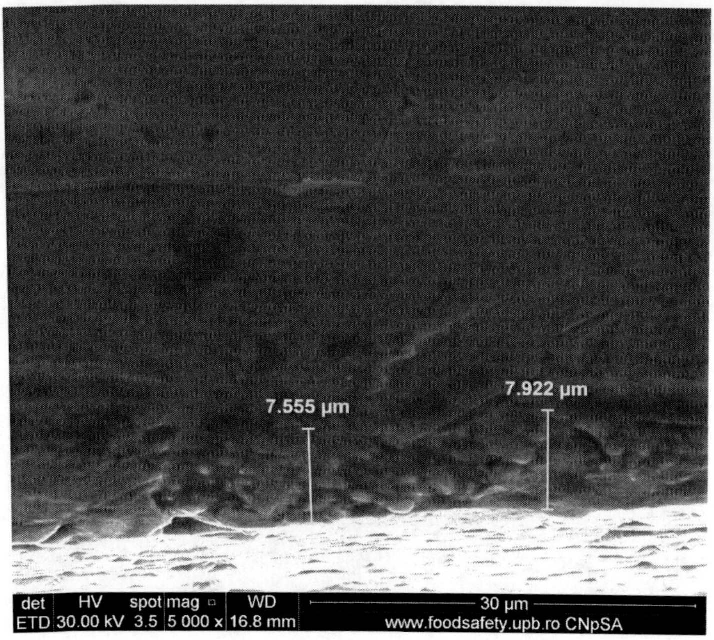

Figure 5.44 Sample from 304 stainless steel with 2 mm thickness cut with laser speed, $v = 80$ mm/s, power, $P = 1000$ W—the cut surface at the laser beam output

Figure 5.45 Sample from C45 steel with 2 mm thickness with cut laser speed, $v = 70$ mm/s, power, $P = 700$ W—general view of cut surface

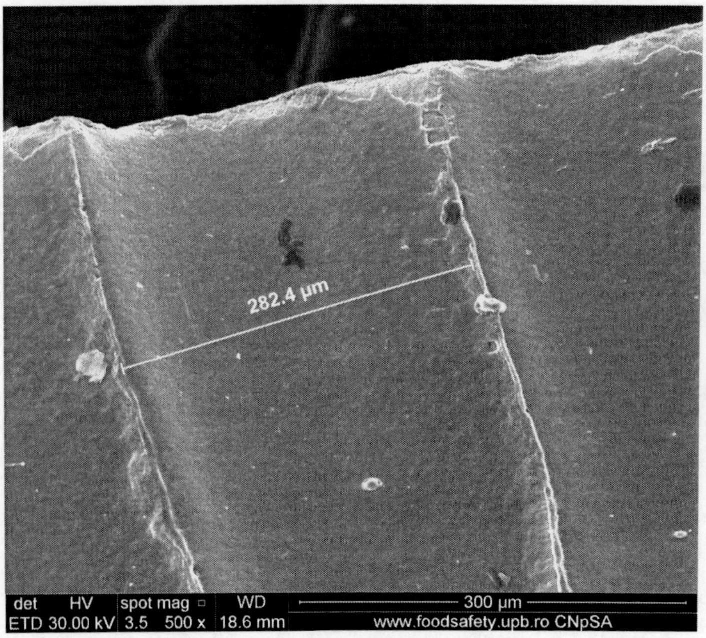

Figure 5.46 Sample from C45 steel with 2 mm thickness with cut laser speed, $v = 70$ mm/s, power, $P = 700$ W—the cut surface at laser beam entrance

Figure 5.47 Sample from C45 steel with 2 mm thickness with cut laser speed, $v = 70$ mm/s, power, $P = 700$ W—the middle of cut surface

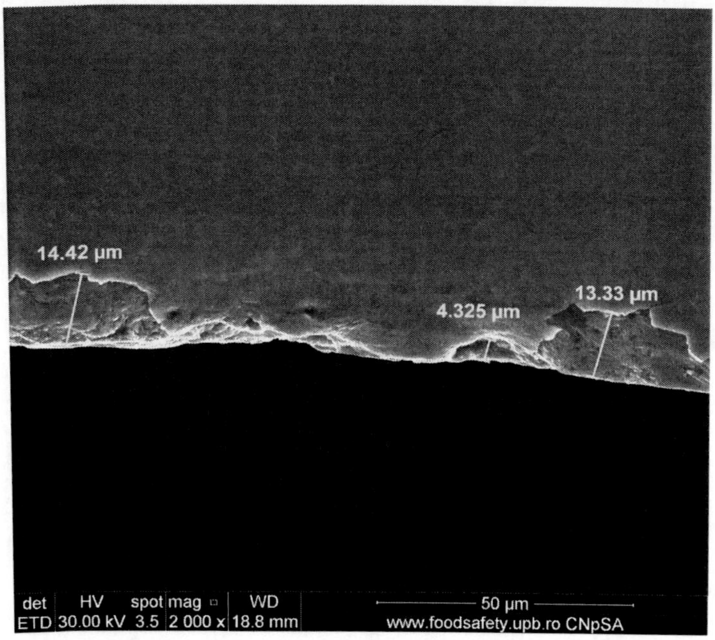

Figure 5.48 Sample from C45 steel with 2 mm thickness cut with laser speed, $v = 70$ mm/s, power, $P = 700$ W—the cut surface at the laser beam output

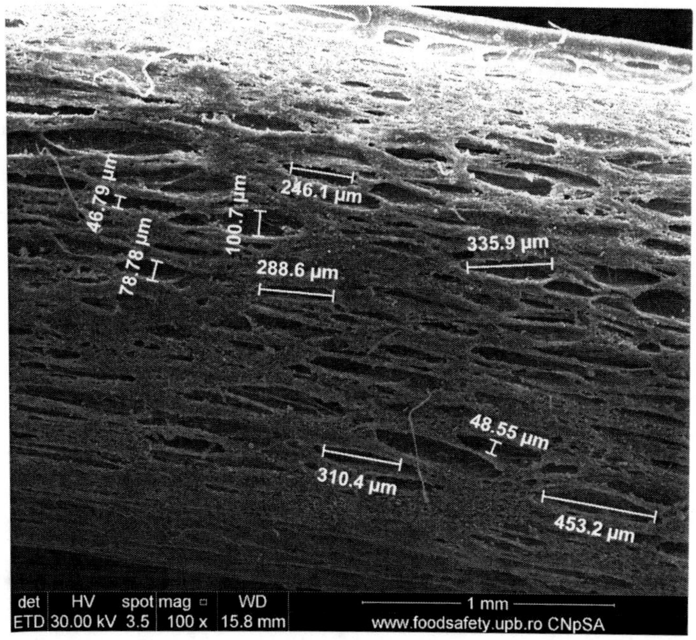

Figure 5.49 Sample from PVC with 2 mm thickness cut with laser speed, $v = 15$ mm/s, power, $P = 78$ W—the general view of cut surface

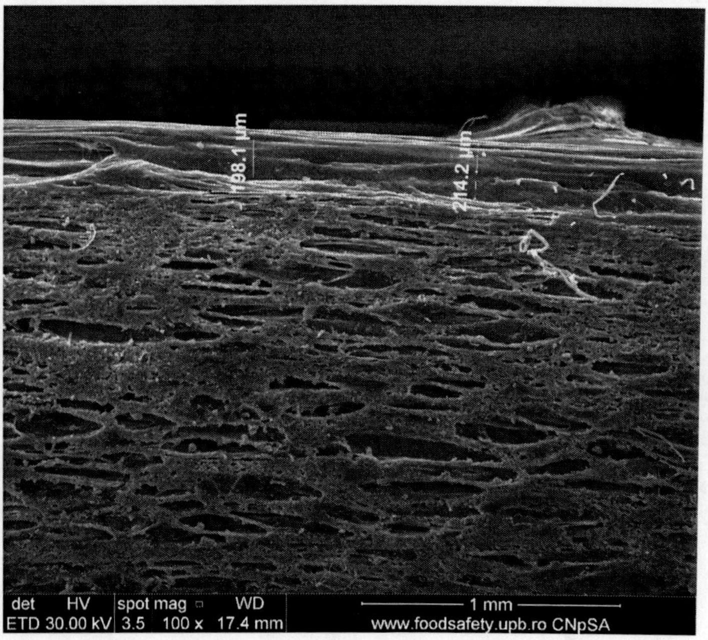

Figure 5.50 Sample from PVC with 2 mm thickness cut with laser speed, $v = 15$ mm/s, power, $P = 78$ W—the cut surface at laser beam entrance

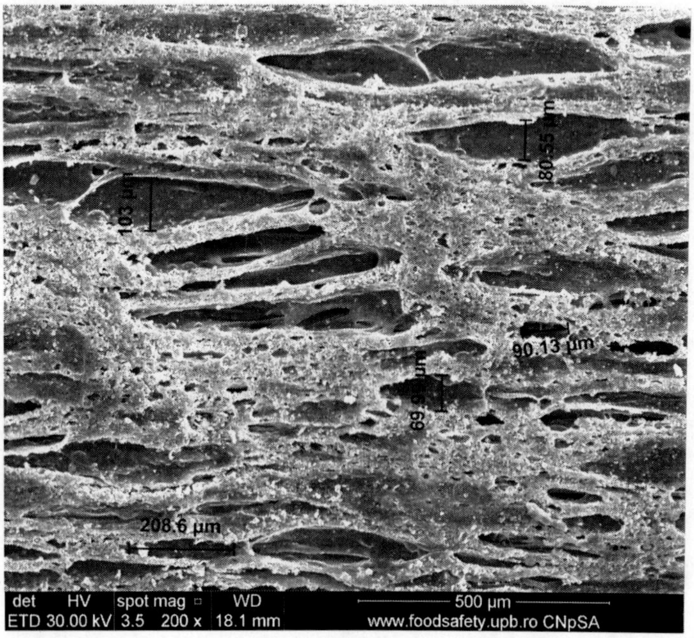

Figure 5.51 Sample from PVC with 2 mm thickness cut with laser speed, $v = 15$ mm/s, power, $P = 78$ W—the middle of cut surface

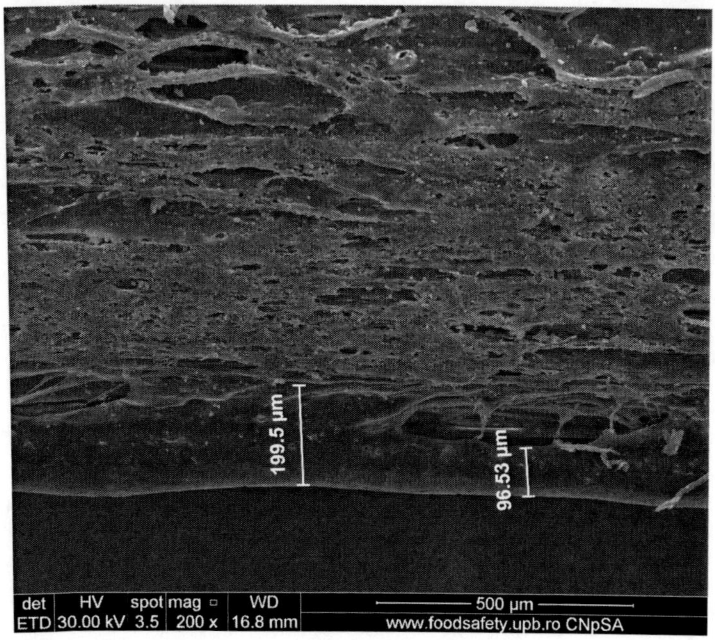

Figure 5.52 Sample from PVC with 2 mm thickness cut with laser speed, $v = 15$ mm/s, power, $P = 78$ W—the cut surface at the laser beam output

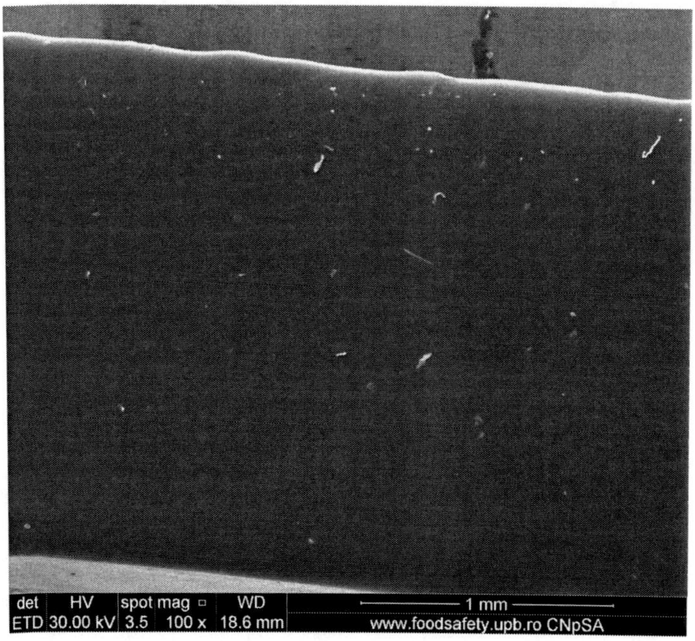

Figure 5.53 Sample from Plexiglas with 2 mm thickness cut with laser speed, $v = 15$ mm/s, power, $P = 78$ W—the general view of cut surface

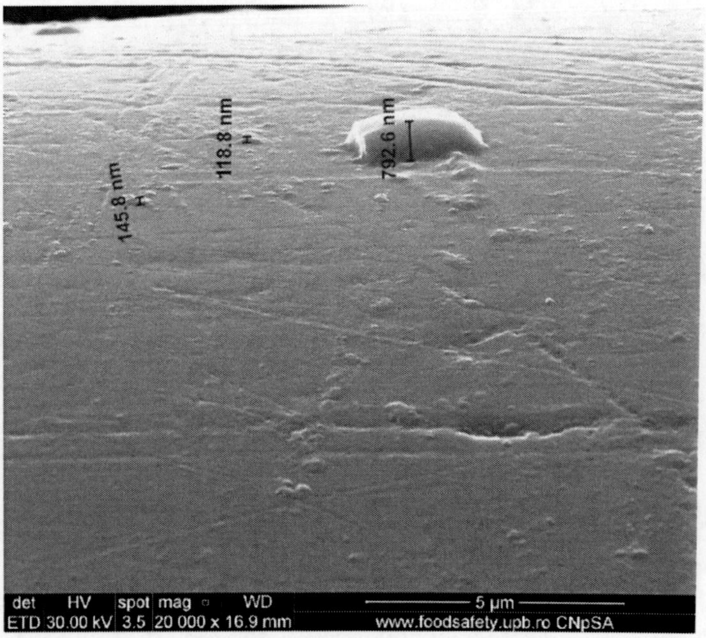

Figure 5.54 Sample from Plexiglas with 2 mm thickness cut with laser speed, $v = 15$ mm/s, power, $P = 78$ W—the cut surface at laser beam entrance

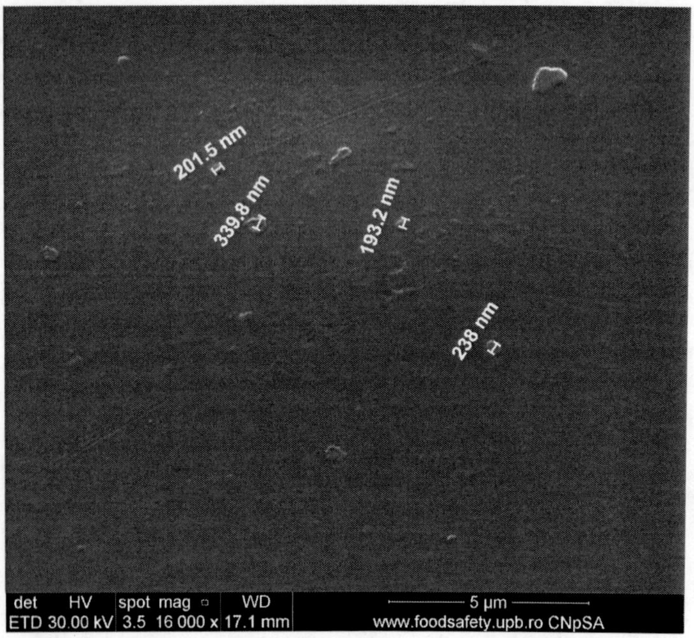

Figure 5.55 Sample from Plexiglas with 2 mm thickness cut with laser speed, $v = 15$ mm/s, power, $P = 78$ W—the middle of cut surface

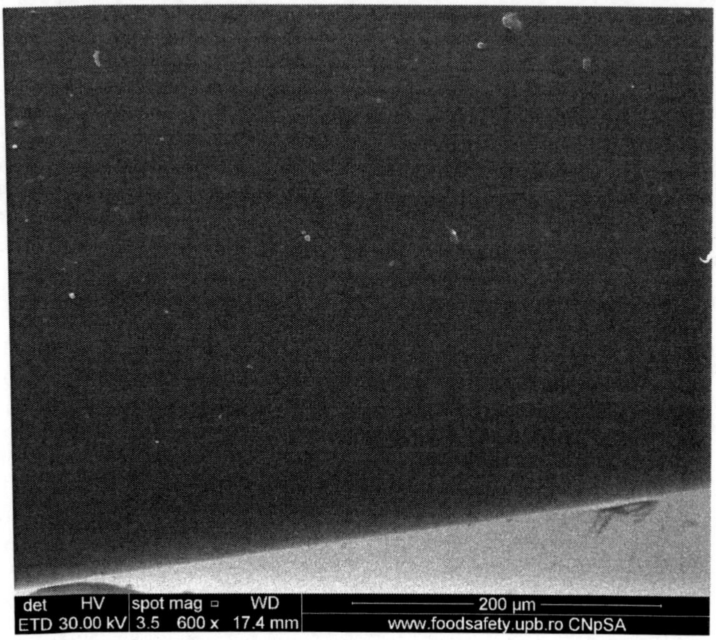

Figure 5.56 Sample from Plexiglas with 2 mm thickness cut with laser speed, $v = 15$ mm/s, power, $P = 78$ W—the cut surface at the laser beam output

Parameters

Name	Expression	Value	Description
dspot	0.015[mm]	1.5E-5 m	laser spot diameter [µm]
gp	2[mm]	0.002 m	sample height
speed	3[m/min]	0.05 m/s	laser speed [m/s]
tstop	(l1-2*rgaz)/speed	0.12 s	cutting time [s]
l1	10[mm]	0.01 m	sample width
l2	20[mm]	0.02 m	sample length
rgaz	2[mm]	0.002 m	gas spot radius
Ttop304	1700	1700	melting point stainless steel 304
P	1000[W]	1000 W	laser power
x0	rgaz	0.002 m	reference distance from laser spot center
sigma	dspot/6	2.5E-6 m	mean square deviation
A1	0.1	0.1	steel absorption coefficient
emi	0.9	0.9	steel emissivity coefficient
h1	10[W/(m^2*K)]	10 W/(m²·K)	heat transfer coefficient
Ed	P/(pi*0.25*dspot^2)	5.6588E12 W/m²	power density on laser spot
xr	x0+speed*tp	0.002 m	mobile reference spot
tp	0[s]	0 s	sweep machining time

Figure 5.57 Parameterization of the model at stainless steel 304 laser cutting. a) the gas spot zone; b) the laser spot zone

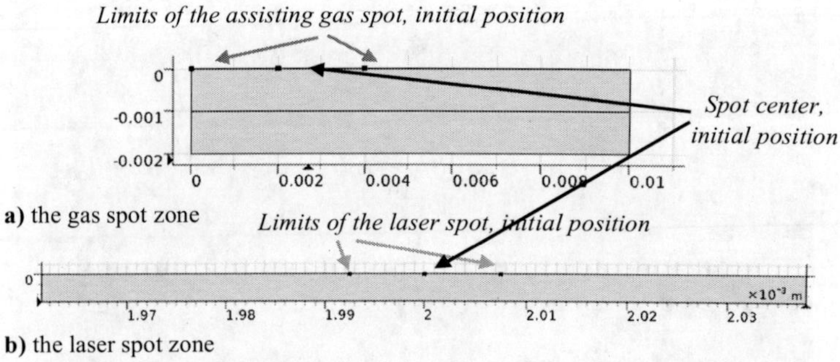

Figure 5.58 Model dynamic geometry

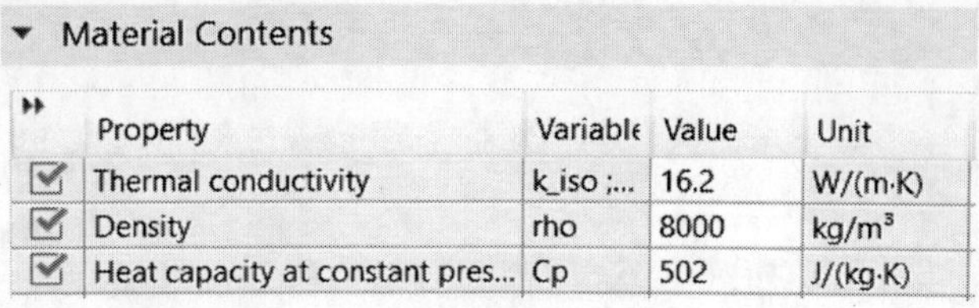

Material Contents

	Property	Variable	Value	Unit
☑	Thermal conductivity	k_iso ;...	16.2	W/(m·K)
☑	Density	rho	8000	kg/m³
☑	Heat capacity at constant pres...	Cp	502	J/(kg·K)

Figure 5.59 Thermo-physical properties of machined material

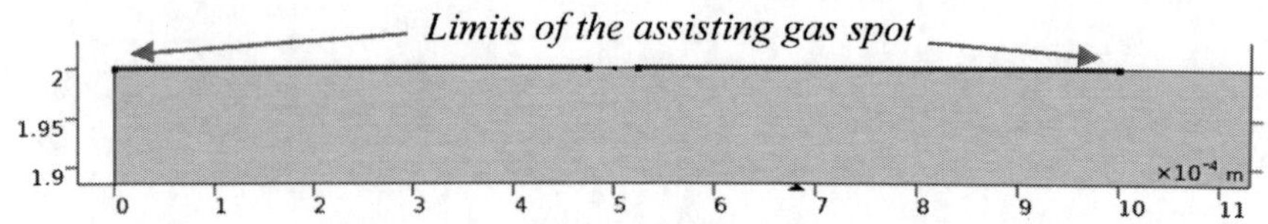

Figure 5.60 Thermal insulation on the surface covered by gas jet

As one can observe the cut surface at Plexiglass has superior quality, very clean with very rare surface damages than the one cut at PVC. The last one has multiple pores with dimensions framed in 0.1–0.2 mm dimensional interval with zone of input and output of laser beam within the material.

The modeling of laser cut process, using Comsol Multiphysics™, Heat Transfer Module, and Stationary comprised the following stages:

a) Parameterization of the model, e.g., laser cutting of stainless steel (Figure 5.57) that facilitates the optimization of the process in terms of energy consumption under conditions of cutting quality achievement.
b) The geometry achievement using built-in tools from Comsol Multiphysics™ (Figure 5.58)
   It can be observed that model geometry is based on a sweep parameter of machining time, i.e., the laser spot is moved (from left to right) across the sample superior surface in order to achieve the cutting on the entire height.
c) Assignment of material properties needed for Heat Transfer Physics—Figure 5.59.
d) Boundary conditions specific to module of heat transfer in solids:
   - the thermal insulation on the surface covered by assisting gas (Figure 5.60),
   - the heat flux on the laser spot (Figure 5.61), related to the variables definition (Figure 5.62),

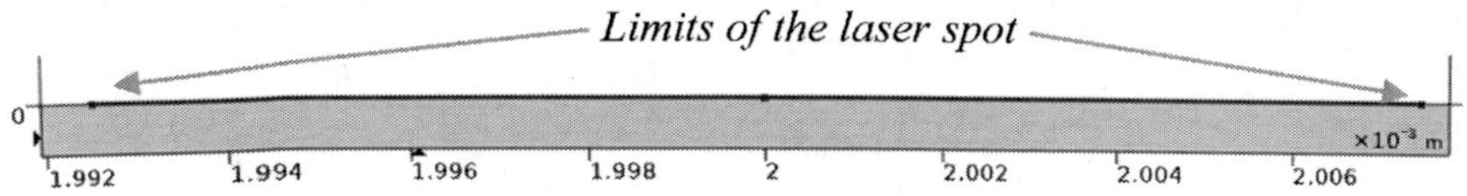

Figure 5.61 Thermal insulation on the surface covered by assisting gas jet

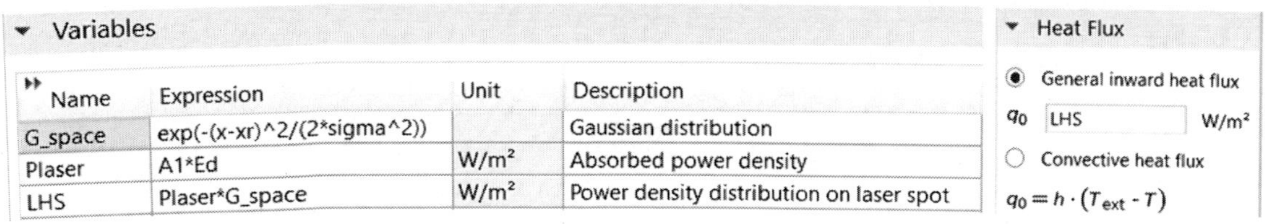
Variables

Name	Expression	Unit	Description
G_space	exp(-(x-xr)^2/(2*sigma^2))		Gaussian distribution
Plaser	A1*Ed	W/m²	Absorbed power density
LHS	Plaser*G_space	W/m²	Power density distribution on laser spot

Figure 5.62 Definition of Gaussian distribution of heat flux on laser spot

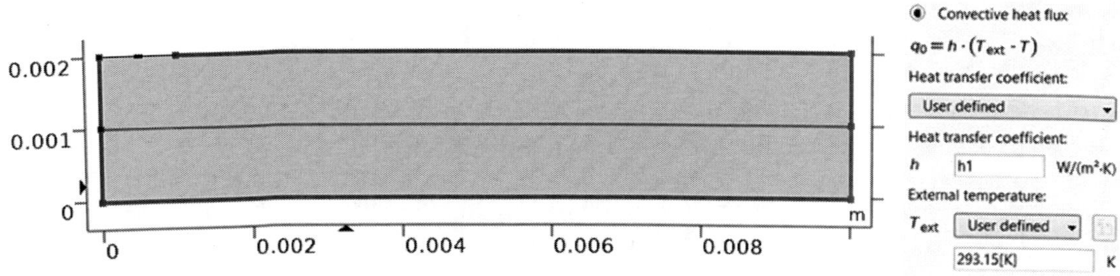

Figure 5.63a Convective cooling condition

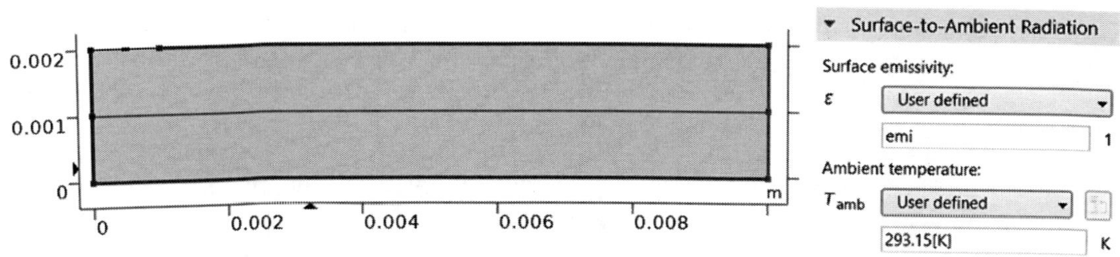

Figure 5.63b Radiation condition

- the convective cooling in air at ambient temperature through the surfaces selected in Figure 5.63a, and by heat transfer coefficient:

The radiation conditions at ambient temperature due to exterior surfaces and emissivity coefficient of the cut material are presented in Figure 5.63b.

The mesh created by free triangular elements and its corresponding statistics are presented in Figure 5.64.

The disposal of thermal front using the sweep parameter of machining time within the cut section of stainless steel 305 is presented in Figure 5.65.

It can be observed that the laser heating produces the cutting on the entire depth of the sample, and the temperatures above melting point indicated the zone of material removed. Two

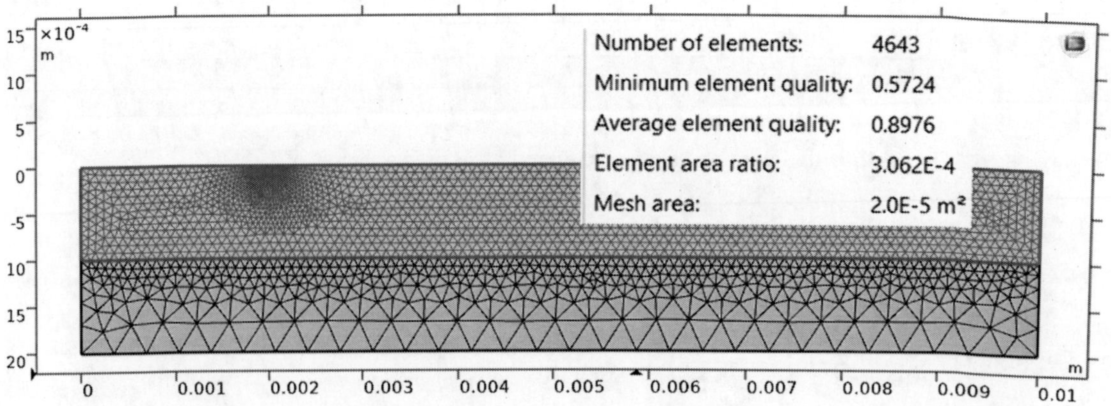

Figure 5.64 The dynamic mesh associated with the sweep parameter

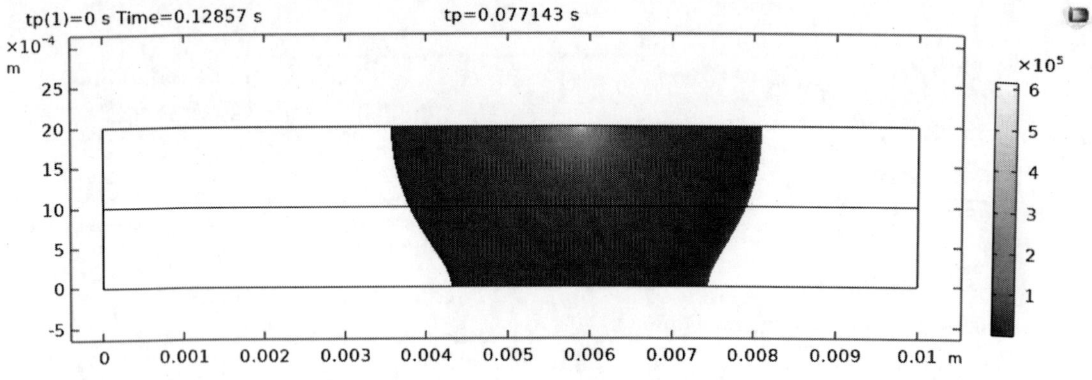

Figure 5.65 The thermal front dynamics during machining at stainless steel 305 laser cutting, spot diameter 0.05 mm, speed 0.05 m/s, power 1,000 W

extreme situations are avoided: (1) the overheating of material is not generated that produces structural damage due to high temperature gradients; (2) partial cutting that produces strips in the cut section, which also reduces the cutting quality.

Similar results were obtained by numerical simulation with the model presented above, in case of C45 cutting, using the following parameters: spot diameter—0.012 mm; cutting speed—90 mm/s; power—1,000 W.

The same stages of modeling were approached with polymeric materials cutting like PVC and Plexiglass.

In particular, some parameters were assigned in this case as it is presented in Figure 5.66.

The material characteristics used for PVC are presented in Figure 5.67.

Numerical simulation of laser cutting with sweep parameter as machining time gave the disposal of the thermal front within the cut section as it is presented in Figure 5.68.

The results obtained are similar to those from steel cutting, aiming at minimization of energy consumption in order to achieve the quality of cutting without streaks, and overheating the material that could lead to structural damages.

Similar results at numerical simulation were obtained at Plexiglas cutting, using a spot diameter of 0.2 mm, cutting speed of 2 mm/s, and power of 30 W.

▾ Parameters

Name	Expression	Value	Description
P	24.6[W]	24.6 W	Laser power
speed	8.877[mm/s]	0.008877 m/s	feed speed [m/s]
dspot	0.12566[mm]	1.2566E-4 m	laser spot diameter
TtopPVC	177+273.15	450.15	melting point of PVC
emi	0.89	0.89	emissivity coefficient
A1	0.28	0.28	absorption coefficient

Figure 5.66 Assignation of some particular parameters at PVC cutting

Property	Variable	Value	Unit
Density	rho	1760[kg/m^3]	kg/m³
Thermal conductivity	k_iso ;...	0.1[W/(m*K)]	W/(m·K)
Heat capacity at constant pressure	Cp	880	J/(kg·K)

Figure 5.67 Allocation of material properties

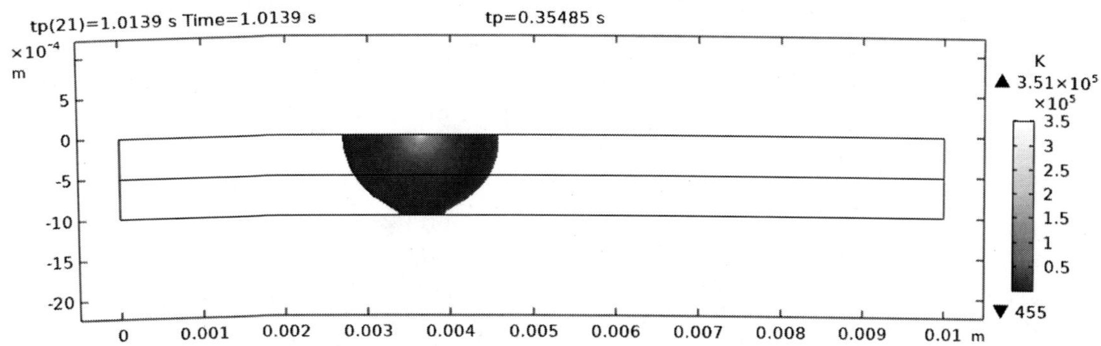

Figure 5.68 Thermal front disposal at PVC laser cutting, spot diameter 0.126 mm, cutting speed 8.877 mm/s, and power 24.6 W

Conclusion

Based on numerical simulation that aimed at cutting optimization, it can be noticed that at steel cutting, like C45 and stainless steel, the power is significantly greater, as well as power density on laser spot, and cutting speed in comparison with the working modes parameters at polymeric materials, like PVC at Plexiglass with the same thickness of samples.

Modeling and Simulation of Plasma Cutting

Plasma cutting is a process in which an inert gas (compressed air) is blown at high speed from a nozzle, and at the same time, an electric arc is formed by a copper electrode at the nozzle level, converting part of the gas into plasma which is focused on the surface to be cut.

Plasma is the fourth state of the matter, a gas-like substance consisting of a mixture of electrons, positive ions, and neutral particles (atoms or molecules) that are in continuous and disordered motion. The concentrations of electrons and ions are approximately equal, so macroscopically the plasma is electrically neutral.

In the cutting process, the plasma arc locally melts the material and removes it at high speed. The high degree of energy concentration and the high temperature of the arc makes it possible to cut metal and metal alloys, high alloy steels, aluminum, copper, titanium, etc. So, Plasma Machining (PM) belongs also to thermal nonconventional machining (Marinescu et al., 2019).

Therefore, both types of machining, LBM and PM, were approached as particular cases of thermal processes by creating a parametric model of material removal, numerically simulated, which will be presented below.

This type of machining was used for cutting some steels like usual C 45 and stainless steel 305 for comparison of technological parameters with those obtained at laser cutting.

The PM parameters are presented in Table 5.12.

The images of plasma cut surfaces of samples from C 45 (Figures 5.69–5.72) and stainless steel 305 (Figures 5.73–5.76) with thickness of 2 mm were taken by QUANTA INSPECT F50 SEM with a resolution of 1 nm.

As one can see from Figures 5.69–5.76, the surface quality is inferior to that of the laser cut on the same grade steel sample. The cut surface is characterized by an input zone of the plasma beam, a central zone with multiple surface damages, and an output zone. Both the input and output zones have a thickness of 0.1 mm order of magnitude. The central zone presents some craters and prominences of 10–30 μm due to the high temperature gradient (Figures 5.71 and 5.75). This was the result of the thermal impact of plasma heating. The working parameters used were: current step, $I = 35$A (power $P = 13.3$ kW), cutting speed $v = 3{,}500$–$4{,}500$ mm/min, and air pressure of 5–6 MPa.

The numerical simulation of plasma cutting comprised similar stages as previous thermal model from laser cutting, using Comsol Multiphysics, Heat Transfer Module, Stationary, and a sweep parameter as machining time. Some particular stages are presented below:

a) Parameterization of the model—Figure 5.77
b) The boundary conditions applied on the geometry of this model in particular were:

The heat flux on the plasma spot is presented in Figure 5.78.

Variables defined related to heat flux on plasma spot that is based on Gauss distribution are presented in Figure 5.79.

Table 5.12 Features of the Powermax105 SYNC Plasma cutting machine

Generator power (kW)	Maximum 30 kW for 105A current
Voltage (V)	380
Current intensity (A)	35–105
Gas supply	Clean, dry air, no oil or oxygen
Optimum intake gas pressure (MPa)	0.76–0.83
Minimum inlet gas pressure (MPa)	0.46
Type of power supply	IGBT inverter (bipolar transistor with isolated deck)

Figure 5.69 Plasma cut surface of C45 steel with 2 mm thickness with cut speed, $v = 0.072$ mm/s, power, $P = 13.3$ kW—general view

Figure 5.70 Plasma cut surface of C45 steel with 2 mm thickness with cut speed, $v = 0.072$ mm/s, power, $P = 13.3$ kW—input zone

Figure 5.71 Plasma cut surface of C45 steel with 2 mm thickness with cut speed, $v = 0.072$ mm/s, power, $P = 13.3$ kW—central zone

Figure 5.72 Plasma cut surface of C45 steel with 2 mm thickness with cut speed, $v = 0.072$ mm/s, power, $P = 13.3$ kW—output zone

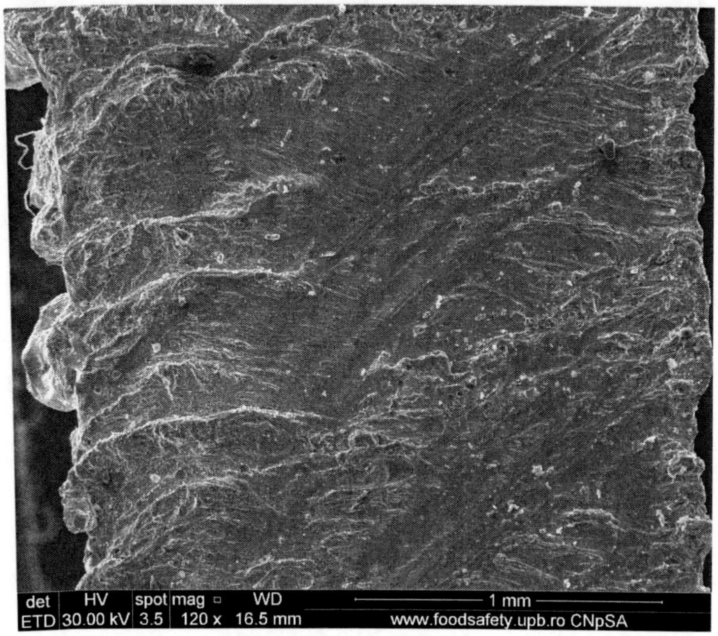

Figure 5.73 Plasma cut surface of stainless steel with 2 mm thickness with cut speed, $v = 0.058$ mm/s, power, $P = 14$ kW—general view

Figure 5.74 Plasma cut surface of stainless steel 2 mm thickness with cut speed, $v = 0.058$ mm/s, power, $P = 14$ kW—input zone

Figure 5.75 Plasma cut surface of stainless steel 2 mm thickness with cut speed, $v = 0.058$ mm/s, power, $P = 14$ kW—central zone

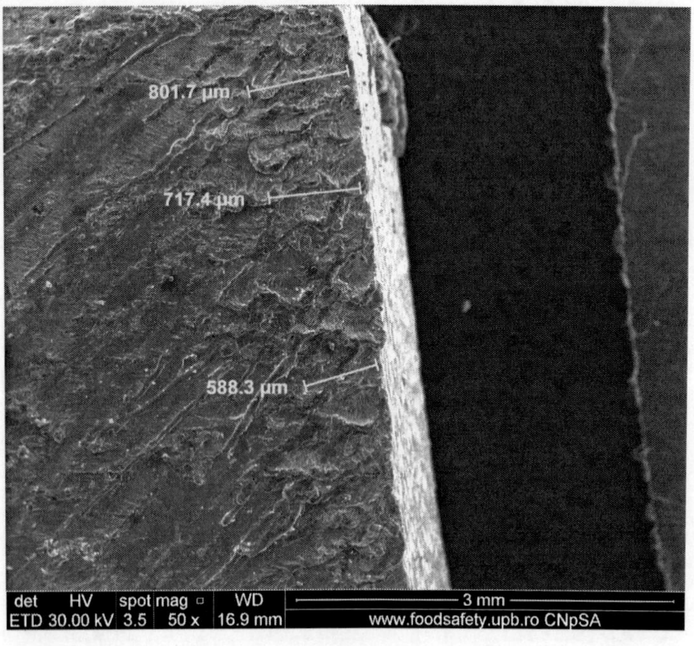

Figure 5.76 Plasma cut surface of stainless steel 2 mm thickness with cut speed, $v = 0.058$ mm/s, power, $P = 14$ kW—output zone

▾ Parameters

Name	Expression	Value	Description
dspot	0.2[mm]	2E-4 m	plasma spot diameter [µm]
P	14000[W]	14000 W	plasma power
Ed	P/(pi*0.25*dspot^2)	4.4563E11 W/m²	power density on plasma spot
sigma	dspot/6	3.3333E-5 m	mean square deviation
speed	5[m/min]	0.083333 m/s	cutting speed [m/s]
tstop	(l1-2*0.2*rgaz)/speed	0.1104 s	machininig time [s]
tp	0[s]	0 s	sweep machining time
l1	10[mm]	0.01 m	sample width
l2	20[mm]	0.02 m	sample length
gp	2[mm]	0.002 m	sample height
xr	x0+speed*tp	2E-4 m	mobile reference spot
x0	rmgaz+dspot/2	2E-4 m	reference distance from laser spot center
rmgaz	0.1[mm]	1E-4 m	radius of gas margin
emi	0.25	0.25	steel emissivity coefficient
A1	0.56	0.56	steel absorption coefficient
h1	10[W/(m^2*K)]	10 W/(m²·K)	heat transfer coefficient
Ttop304	1400+273.15	1673.2	melting point of stainless steel 304

Figure 5.77 Parameters of plasma cutting model

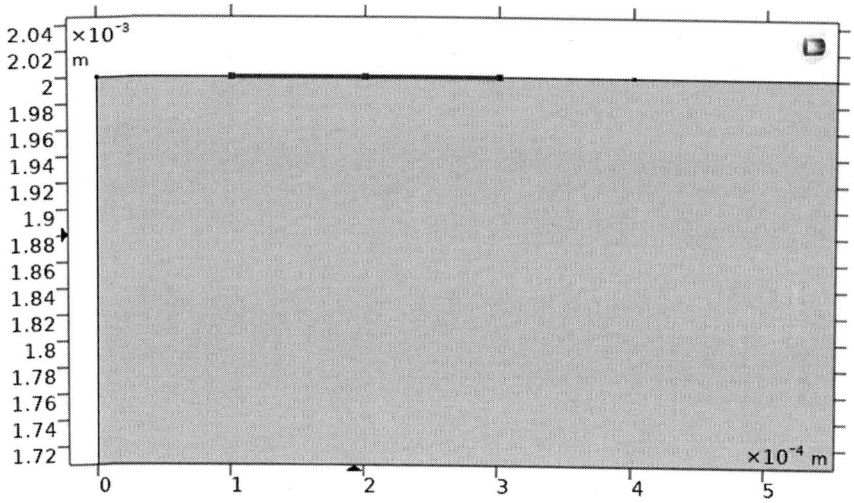

Figure 5.78 Plasma spot geometry

▾ Variables

Name	Expression	Unit	Description
G_space	exp(-(x-xr)^2/(2*sigma^2))		Gauss distribution of energy on plasma spot
Pplasma	A1*Ed	W/m²	power density absorbed
PHS	Pplasma*G_space	W/m²	power density distribution on plasma spot

Figure 5.79 Variables defined for heat flux on plasma spot

Thermal insulation produced by the remaining gas surrounding the plasma jet is shown in Figure 5.80.

The boundary conditions of cooling and radiation are similar to those from the previous laser cutting model, presented above, referring to the surfaces in contact with the ambient working environment.

c) Solving the model—The results obtained in terms of thermal front disposal during the plasma cutting process are presented in Figure 5.81.

The used cutting parameters provide appropriate heating for surface quality achievement, without overheating and underheating that could lead to structural material damages or incomplete cutting. As one can see the working parameters from the numerical simulation are in agreement with the real experimental data, validating the model.

Other similar results that could determine machined surface quality were obtained by numerical simulation of plasma cutting of C45 steel, using a spot diameter of 0.2 mm, cutting speed of 0.07 mm/s, and power of 1.4 kW.

Conclusion

Numerical simulation of plasma cutting of some steels like C45 and stainless steel 305 pointed out that for the increase of quality of machined surface, the plasma power is much greater than in the case of laser cutting (same materials comparison), the spot diameter is greater, but the cutting speed is decreased.

Modeling and Simulation of Electrochemical Polishing

Over time, the performance of technological equipment has evolved, allowing parts to be obtained as accurately as possible and with a lower manufacturing cost. At the same time, the complexity of the parts increased, which led to improved processes used, obtaining hybrid

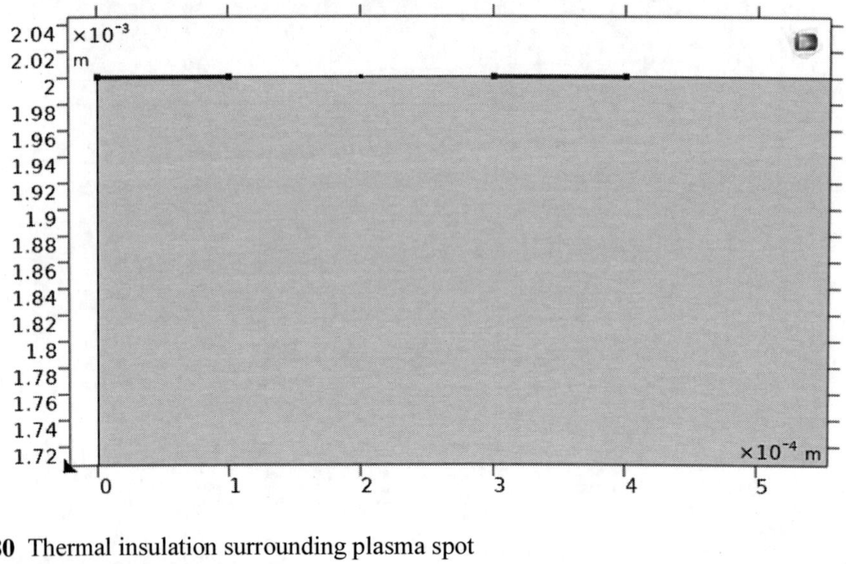

Figure 5.80 Thermal insulation surrounding plasma spot

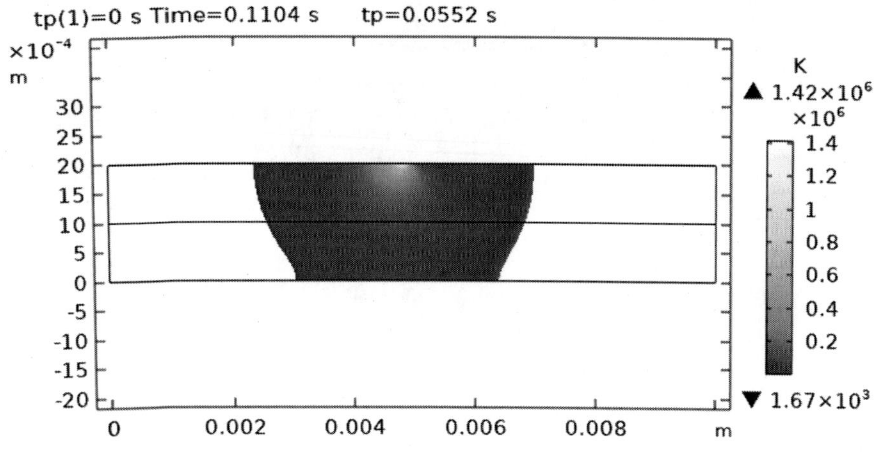

Figure 5.81 The thermal front dynamics during machining at stainless steel 305 plasma cutting, spot diameter 0.2 mm, speed 0.083 m/s, power 14,000 W

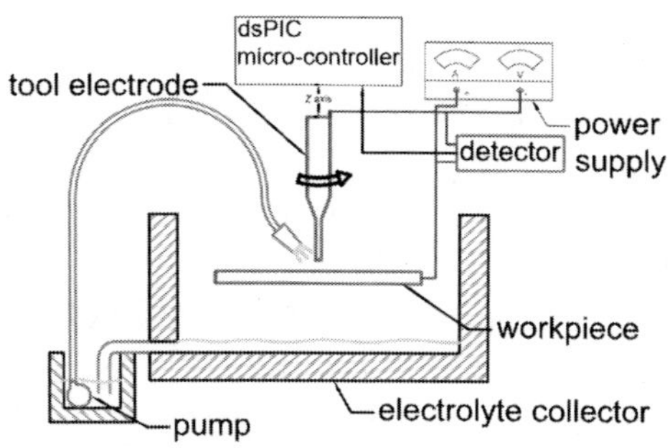

Figure 5.82 Electrochemical process diagram

procedures and developing new tools and equipment. Electrochemical polishing is the process of removing a uniform superficial layer from metal parts to improve their corrosion resistance, as well as improve the performance and life cycle of parts. Many companies use electrochemical polishing as a "final step" in the manufacturing process. The process of Electrochemical Machining (ECM) in different variants like 3D machining, surface polishing, deburring, drilling, and also many other similar classic machining as turning, milling, drilling, honing, etc. can be grouped into electric nonconventional machining, from the criterion of energy used directly in the material removal mechanism (Ghiculescu, 2004).

This study aims to understand the phenomena regarding laminar flow and current distribution based on numerical simulations, components of ECM polishing, using dedicated software, and Comsol Multiphysics.

Electrochemical polishing, also known as anodic polishing or electrolytic polishing, is an electrochemical process that removes material from a metal part, reducing surface roughness by leveling micro-peaks and improving surface finish, as Figure 5.82 presents. This process is the inverse of galvanizing.

It can be used instead of fine abrasive polishing in the microstructural preparation of surfaces. Based on the combination of electricity and chemistry, ECM was used to improve the surface quality of the electrically materials, i.e., metals. Therefore, it turns into a popular finishing in a wide variety of industries, including the following (Figure 5.83):

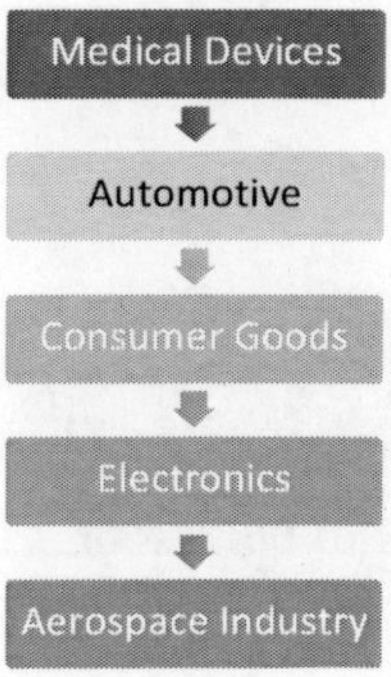

Figure 5.83 ECM polishing applications

Some Advantages of Electrochemical Polishing Process

- One of the advantages is the decorative aspect; the finish is glossy, often very reflective.
- Microscopic smoothness can be another advantage. Electrochemical polishing levels the peaks and valleys and can decrease the surface roughness (Ra) up to 50%.
- At the microscopic level, electrochemical polishing removes small volumes of material from the surface that makes it appropriate for micromachining.
- Electrochemical polishing provides an ultra-clean surface that contributes to sterilization and diminish corrosion.
- Applicable to fragile and complex parts, which could be damaged during manipulation; thus, it can replace some other mechanical finishing techniques.
- Can polish inaccessible areas by mechanical polishing methods; effectively machining only small areas on complex parts.

Basic Phenomena in Electrochemical Polishing

Electrochemical polishing (superfinishing, ECM) of some parts of Fe, Co, Ni, Cr, Al, etc. and their alloys involves the making of an anode consisting of the object to be processed and a cathode represented by the tool in an electrolyte bath containing basic, acidic, or neutral solutions containing ions which occupy a position in the liotropic series (according to Cooper) after the sulfate ion. These ions can form easily, soluble salts with the abovementioned metals, using a relatively low voltage (less than 24 V), current density ($J = 0.05–0.5$ A/cm^2) enough to produce the removal oxidation products formed on the surface of the anode.

Electrochemical polishing reduces the roughness of the processed surface, according to the mechanism shown in Figure 5.84, based on the phenomenon of anodic dissolution, which occurs in the electrolytic environment (1) because of the electric field E created between the tool (2) (cathode) and the part (3) (anode). On the surface of the workpiece, a passivated layer (4) (viscous, electrically neutral film) is formed from the dissolved products, the thickness of which is smaller in the region of the micropeaks, from which it results that its electrical resistance is lower in these areas. Moreover, the intensity E is higher in the micropeak area (Ghiculescu, 2004).

The electrolyte flow rate exerts a significant influence on polishing rate. Some experimental results show that for a specific configuration of the microgeometry, the polishing rate increases

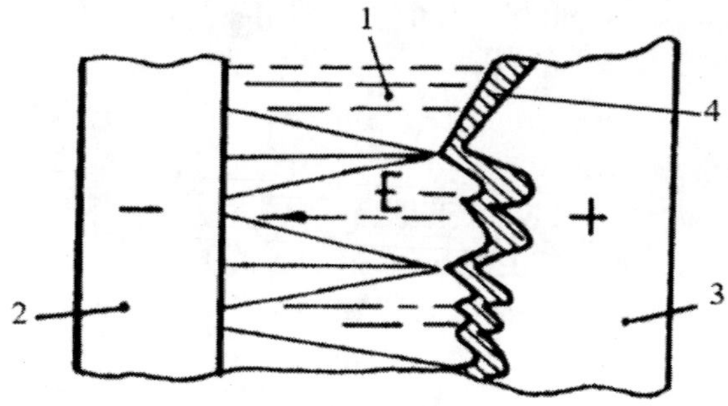

Figure 5.84 Diagram of the ECM polishing mechanism

to a certain flow rate, after which it decreases to higher flow rates. Theoretical models do not provide a critical flow rate for maximum polishing speed since those models were not designed to account for the effect of dynamic flow.

Theories in the literature show that when micro and macro vortices are formed, the rate of anodic dissolution depends on the amount of products resulting from the processing, which are transported to the mainstream. Whenever the materials resulting from the process are retained by the vortices formed at the micro- and macrogeometric level and are not transported to the main current, the ratio will become high and therefore the polishing speed will be lower (Marinescu and Ghiculescu, 2005).

Tool Design in Electrochemical Polishing

Although special attention has been paid to the design of the cathode device when introducing ECMs into manufacturing assembly companies, the configuration of the device remains a remarkable test. The shape of the tool is rarely an ideal negative image of the workpiece due to numerous perturbing factors during the process. Therefore, before tool execution, finite element modeling of ECM process is required.

Stages of Modeling in Electrochemical Polishing

Several stages of modeling in Comsol Multiphysics are presented for the following components of ECM process.

Setting the Working Conditions: Physics and Modeling Space

For this specific modeling, the working space was the 2D *Axisymmetric*, then the Physics module used: Fluid Flow-Laminar Flow at stationary regime. This assumes that the electrolyte flow is laminar, based on Reynolds number as a criterion (Janna, 2020):

$$\mathrm{Re} = \rho\, u\, D / \mu \tag{5.3}$$

where ρ—electrolyte solution density (kg/m^3); u—flow velocity of electrolyte (m/s); D—transversal dimension of the flowing path, respective the interelectrode gap (m); μ—dynamic viscosity (Pa·s). In this case, Re is 0.3, much smaller than the threshold between laminar and turbulent flow, 2,100.

Parameterization of the Model

The parameters considered were those related to the geometry of the part to be polished, the tool, the surface roughness, the electrochemical characteristics of electrolyte solution, materials of the tool, and the machined part. In Figure 5.85, the parameterization of the model is presented.

The Geometry of the Model

The geometry was created using built-in tools from Comsol Multiphysics, addressing the working zone (Figure 5.86). The microgeometry was assimilated by a series of ellipses, based on Rz roughness, and its pitch, larger than Rz value.

The Allocation of the Material

The allocation of the material in the Flowing module is presented in Figure 5.87, along with the Material contents window. Thus, the fluid is assimilated with water, because the proportion of NaCl in solution is over 90%. The machined part is from X210Cr12 steel (D3 or UNS T30403), and the material assigned for the tool is copper 99. Their characteristics were considered in the boundary conditions, specific to the Secondary current distribution model.

The Boundary Conditions

The boundary conditions for this study are shown in Figure 5.88, in the form of an inlet and outlet at atmospheric pressure, Electrolyte inlet velocity, *Vel 1* (m/s); Outlet pressure (atmospheric pressure), *Pout* (Pa); Wall condition is no slip.

The boundary conditions related to the secondary distribution current are presented in Figures 5.89–5.91.

▾ Parameters

Name	Expression	Value	Description
W1	10[mm]	0.01 m	Radius (part)
H1	0.4[mm]	4E-4 m	Working gap
W2	5[mm]	0.005 m	Radius (tool)
H2	1.5[mm]	0.0015 m	Heigh of the electrolyte channel
a	0.0001[m]	1E-4 m	iteration
b	0.000032[m]	3.2E-5 m	Rz (roughness)
sigma	7.95[S/m]	7.95 S/m	Conductivity for the NaCl electrolyte
Eeq_K	-0.85[V]	-0.85 V	Equilibrium potential for Cathode
Eeq_A	-1.55[V]	-1.55 V	Equilibrium potential for Anode
i0_K	0.3[A/m^2]	0.3 A/m^2	Cathode exchange current density
i0_A	1[A/m^2]	1 A/m^2	Anode exchange current density
b_K	-220[mV]	-0.22 V	Tafel curve, cathode
b_A	55[mV]	0.055 V	Tafel curve, anode

Figure 5.85 Parameters of the model

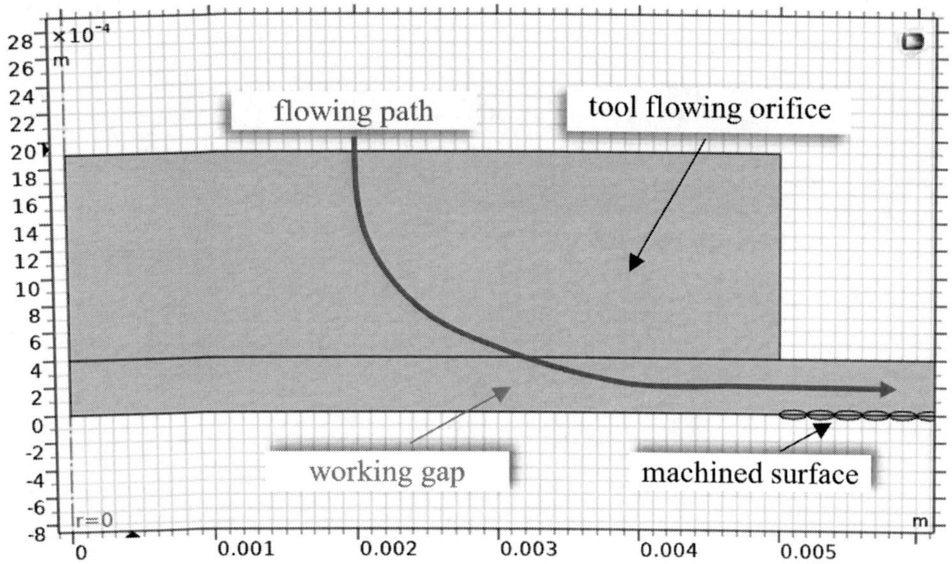

Figure 5.86 The geometry of ECM polishing model

	Property	Varial	Value	Unit
☑	Density	rho	rho_liquid_2(T[1/K])[kg/m^3]	kg/m³
☑	Dynamic viscosity	mu	eta_liquid_1(T[1/K])[Pa*s]	Pa·s

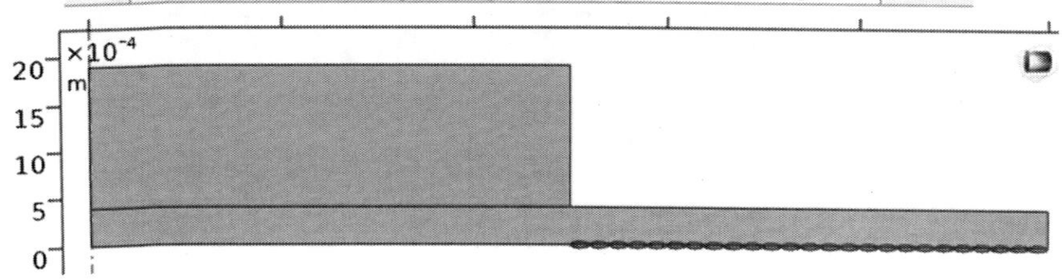

Figure 5.87 Material assignment in the Flowing module

Finite Element Discretization

The discretization is created with triangular elements of finer size, and its quality on 0–1 scale (up) and statistics (down) are presented in Figure 5.92.

The Simulation

Running the model in stationary mode provides the results, in terms of fluid flow velocity variation (Figure 5.93) and pressure variation (Figure 5.94), indicating unavoidable loss of pressure along the flowing path.

The variation of flow velocity pointed out that on the peaks of the microgeometry, the values are greater than in the valleys, which confirm that material is removed chiefly from the prominence zones, contributing thus to the surface roughness decrease, that is, polishing. Moreover, by reducing the frontal working gap (*sf*), the flow velocity grows, which shows an increase in the machining rate, and efficiency of the process—Figure 5.95.

The distribution of the electrolyte potential within the working gap is presented in Figure 5.96.

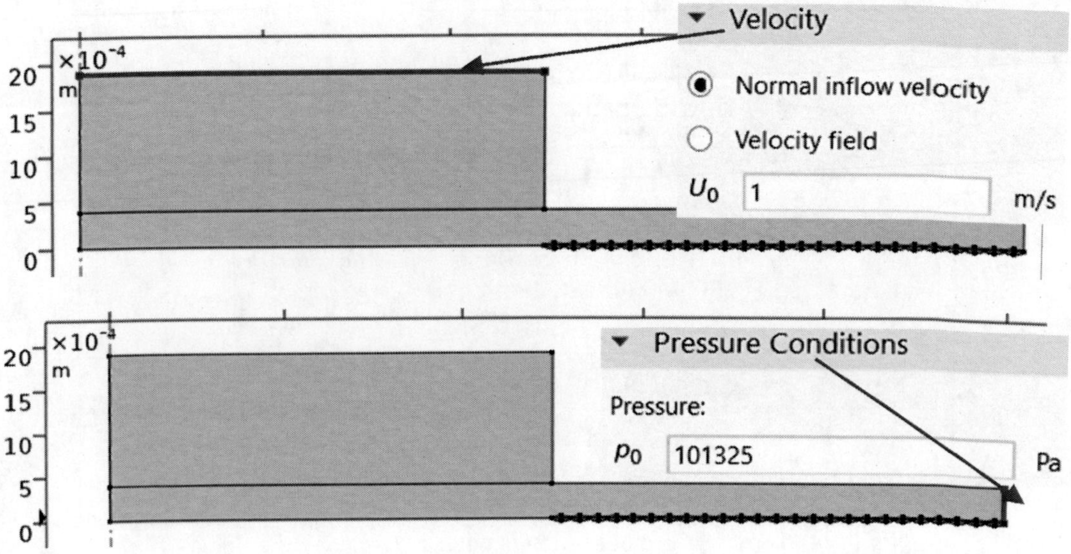

Figure 5.88 Boundary conditions

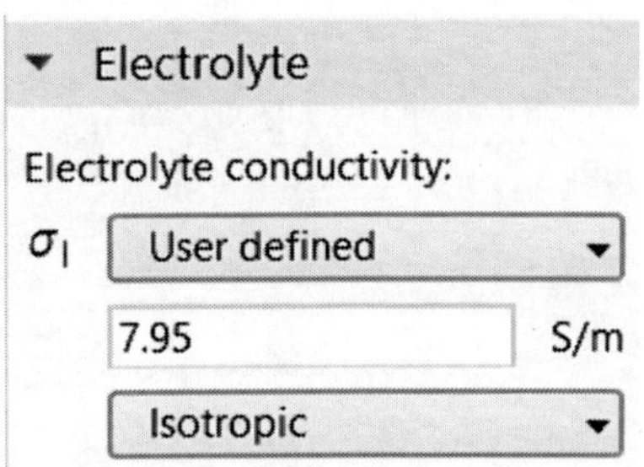

Figure 5.89 Boundary characteristics of NaCl electrolyte solution

It can be noticed that the electrolyte potential grows within the working gap, compared to the flowing orifice from the tool-electrode, which emphasizes that the machining rate is greater on the polished surface (Figure 5.96). Furthermore, electrolyte current density vector (streamlines) shows an increase in the peaks of the microgeometry, compared to valleys, which emphasizes that material removal takes place with priority from these zones, contributing to the polishing process (detail, Figure 5.96). Additionally, from Figure 5.97, when frontal working gap decreases, the electrolyte potential grows, which accelerates the material removal process.

Conclusion

The ECM polishing process was analyzed in terms of electrolyte flow velocity and electrolyte potential. The process numerical simulation confirms the theoretical and experimental data, through which, the material removal rate is greater on the peaks of the microgeometry of the machined surface. Thus, rapid decrease of surface roughness is produced, i.e., high machining rate, under conditions of other numerous advantages of electrochemical machining.

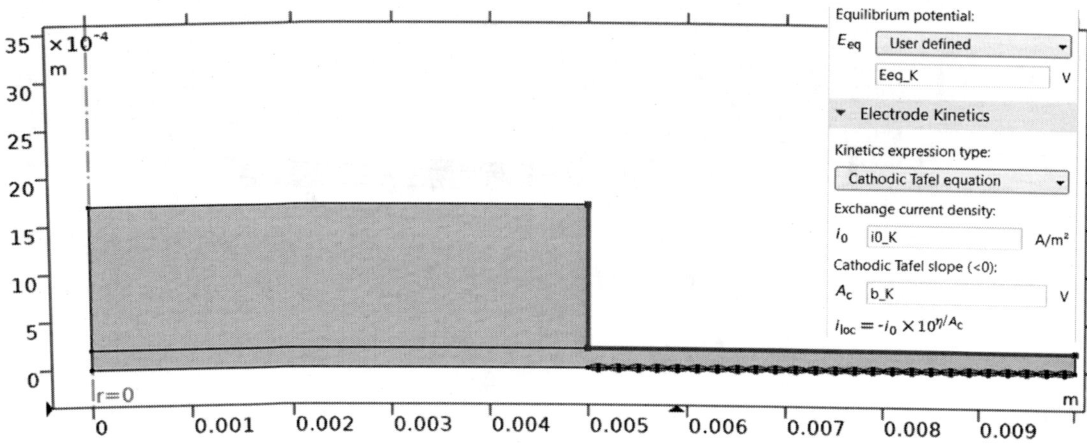

Figure 5.90 Characteristics of electrode surface 1

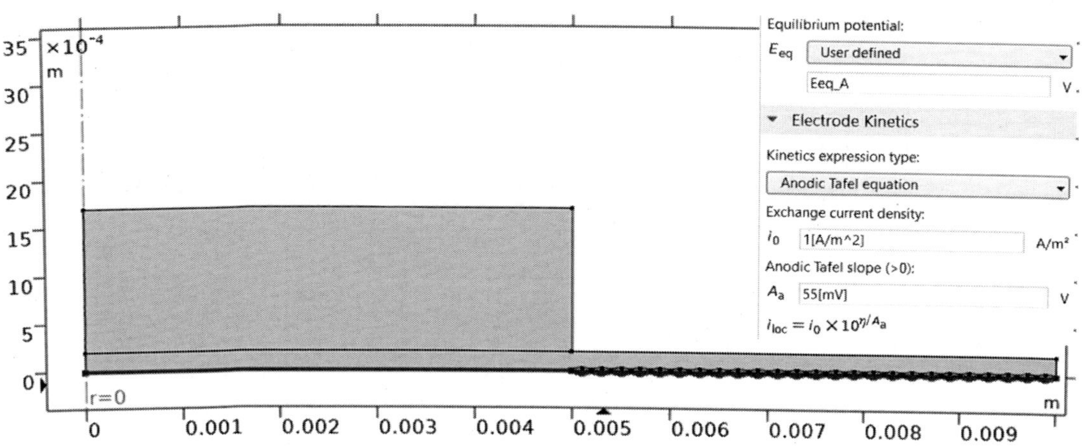

Figure 5.91 Characteristics of electrode surface 2

Modeling of Electrical Discharge Deposition

The need for miniaturization of parts is growing day by day and this market demand has led to the development of many technologies. Some of them are based on the bottom-up approach by growing parts layer by layer, thus obtaining more complex parts in a single configuration. This is explained by microparticles ability to move and fuse with an energy source needed to deposit the material (Jitendra et al., 2010).

A new technology developed in this frame, becoming important in the research communities, is EDD (Electrical Discharge Deposition). EDM (Electrical Discharge Machining) and EDD work on the same principle, but the polarity changes. By using electrical discharge deposition to melt or gasify the electrode, the material is removed from the tool and then moves in the discharge channel, as the material cools over the surface of the workpiece (Zilong and Guan, 2010). The deposition rate is higher than the removal rate, i.e., the wear rate of the tool electrode is higher in the EDD process. The cathode surface will emit a large number of electrons when the surface of the cathode and the anode are close enough (gap distance μm). Those electrons

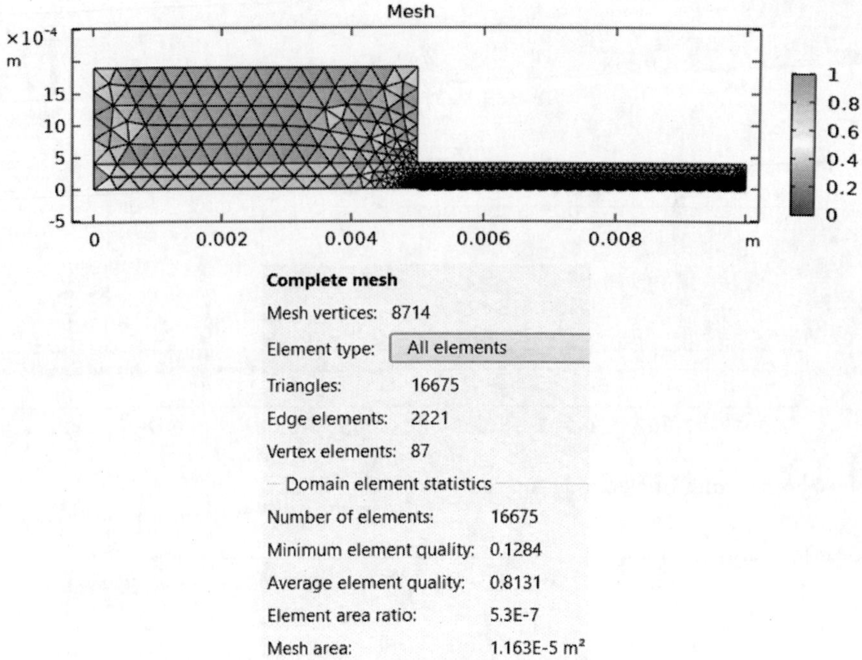

Figure 5.92 Mesh of the model and its characteristics a) Flow velocity distribution—general view; b) Flow velocity distribution—detail

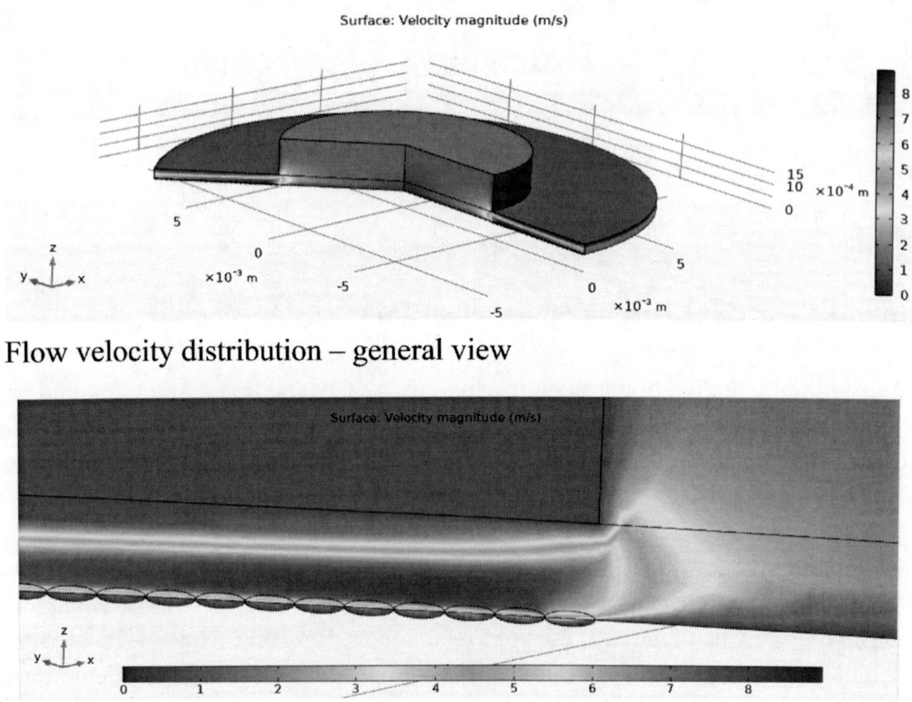

Figure 5.93 Variation of flow velocity within the working zone

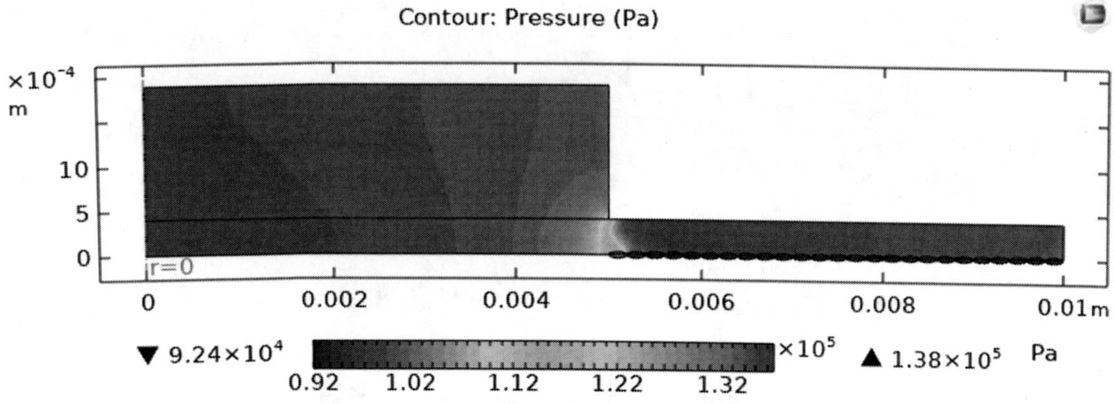

Figure 5.94 Variation of the pressure of the electrolyte along the flow path

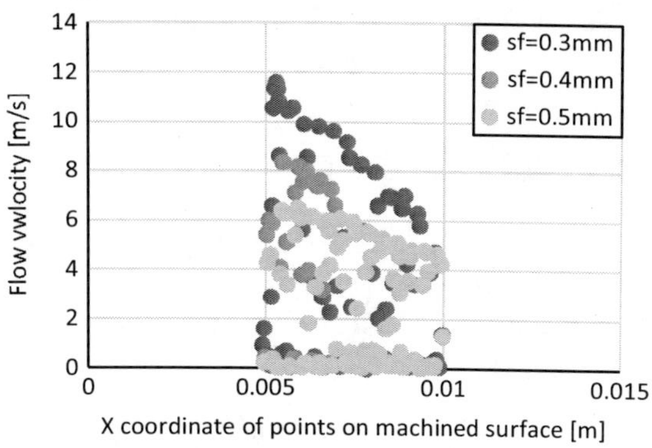

Figure 5.95 Variation of flow velocity in respect of frontal working gap. a) General view of electrolyte potential distribution; b) Detail of electrolyte potential distribution on the machined surface

are accelerated and go to the anode due to the force of the electric field. Collisions between high-velocity electrons and medium-sized dielectric particles create many electropositive particles. During discharge, the electric particles collide continuously, thus forming a plasma channel (Wang et al., 2007).

As shown in Figure 5.98, the tool is linked to the positive pole, and the workpiece to the negative one. The dielectric used in EDM processes is normally used for generating plasma between electrodes by concentrating energy transfer from tool to workpiece and for flushing away molten material. The dielectric preferred for μ-EDD is always air since no material from the workpiece needs to be removed. A tool in EDD process is usually connected to a positive pole from which heavy ions emit from the tool surface and a workpiece, lighter electrons emit from the workpiece that is more numerous than the heavier ions. Because ions are heavier than the workpiece surface, they deposit tool material on it (Jitendra et al., 2010). Therefore considering, that in this case, the power allocated to the tool has to be maximized, when using positive polarity, the duration of the discharge has to be relatively short (Ghiculescu, 2004).

When using an EDM machine (Figure 5.99), the following factors were varied: current, voltage, pulse time, and pause time.

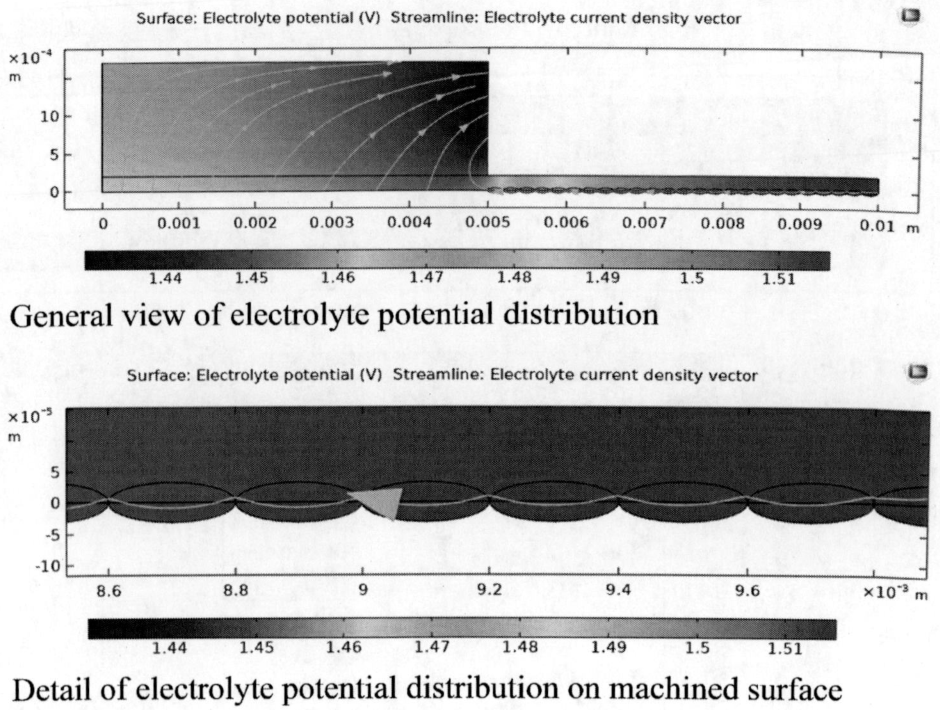

Figure 5.96 Variation of electrolyte potential along the flowing path

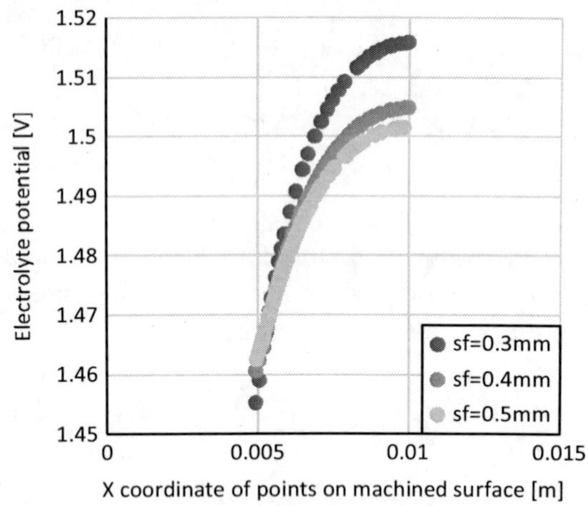

Figure 5.97 Electrolyte potential variation in respect of frontal working gap

The EDD process can be aided by magnetic field (process hybridization) aiming at quality improvement of deposition, in terms of deposit dimensions: height, width, and weight. The variation of the parameters specific to EDM and magnetic flux varies as follows: current (A), $I = 3$, 9, 14, 25; voltage (V); $U = 2$, 30, 50, 70, 98; pulse time (μs), $t_i = 7.5$, 20, 30, 40, 50, pause time, t_0 (μs)=3, 4, magnetic flux (T), $B = 0.07$, 0.1, 0.25, 0.3, 0.4 (Muralidharan et al., 2015).

The materials like tungsten, steel, and brass were deposited on a high-speed steel surface. The results indicated that the deposited structure and the tool electrode are similar, and there is

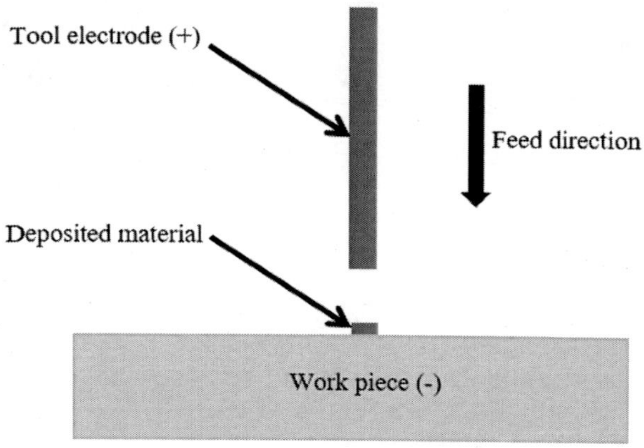

Figure 5.98 Block diagram of electrical discharge deposition (EDD)

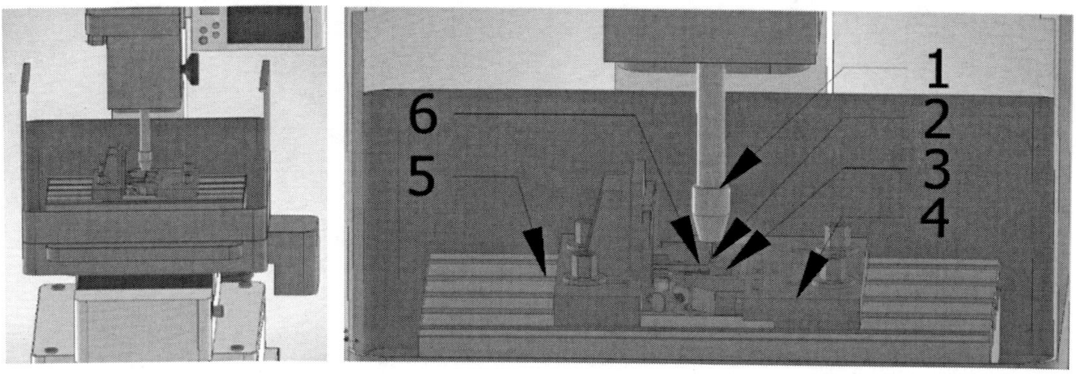

Figure 5.99 Components of technological system of EDD (left—EDM machine, right—detail); 1. holder; 2. tool-electrode; 3. workpiece; 4. workpiece clamping device; 5. worktable; 6. electromagnet clamping device

an interface between the deposited structure and the substrate (Jitendra et al., 2010). The effects of different μ-EDD parameters have not been systematically studied yet.

The aim of this study was to analyze the influence of magnetic field on EDD process and to improve the quality of deposition of tool-electrode conductive materials over the workpiece substrate by using reverse EDM.

For this purpose, Comsol Multiphysics™ was used to model and simulate the ion trajectory in the electromagnetic field using the following settings: 3D - AC/DC - Magnetic Fields - Stationary, in stage 1, as well as Charged Particle Tracing module, in stage 2.

The model was parameterized as it is presented in Figure 5.100, comprising the major elements: anode, coil, and workspace (equivalent of collimator or column). The sample was placed at the bottom of the workspace (Figure 5.101). The process of copper deposition on a sample from steel was considered.

The materials for the model were allocated: air for the workspace, that is, column (Figure 5.102) and copper for the coil and anode tool (Figure 5.103).

Figure 5.104 presents the quality of the mesh, based on tetrahedral elements (choosing the finer grade) on a scale of 0–1, and its statistics.

▾ Parameters

Name	Expression	Value	Description
Ic	0.4[A]	0.4 A	current in the coil
Nc	3000	3000	number of turns
rbe	8	8	outer radius coil
hb	3	3	coil height
rbi	4	4	inner radius coil
rcol	12.5	12.5	column radius
hcol	13	13	column height
scol	3	3	target distance in the column
hc	4	4	cathode height
rc	0.5	0.5	cathode radius
dc	4	4	distance cathode_coil
E0	500[eV]	8.0109E-17 J	initial energy
mCu	63.546*1.66053906660e-27	1.0552E-25	copper ion mass
V0	9.7349e13	9.7349E13	initial speed of copper ion
Jsz	500000	5E5	surface current density
Np	1000	1000	number of particles
VCu	1.3*10e7	1.3E8	copper ion velocity
tpN	scol/VCu	2.3077E-8	copper ion travel time
tps	10e-4	0.001	copper ion study time

Figure 5.100 Parameters of the electrical discharge deposition model

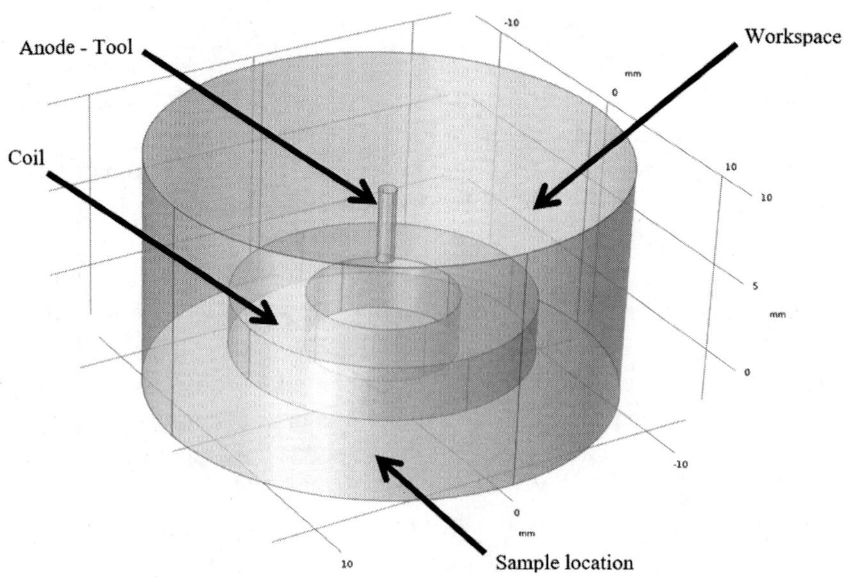

Figure 5.101 Geometry of the model (mm)

Property	Variable	Value	Unit
Electrical conductivity	sigma_iso ; sigmaii = sigma_iso, sigmaij = 0	0[S/m]	S/m
Relative permeability	mur_iso ; murii = mur_iso, murij = 0	1	1
Relative permittivity	epsilonr_iso ; epsilonrii = epsilonr_iso, epsilonrij = 0	1	1

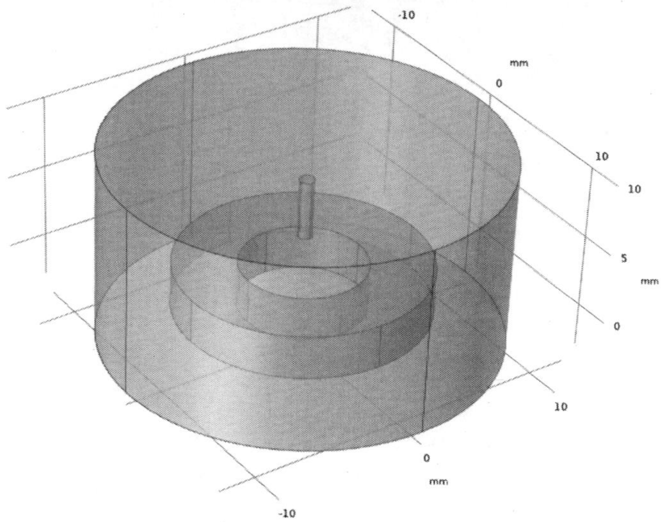

Figure 5.102 Material allocation for the column

Property	Variable	Value	Unit
Relative permeability	mur_iso ; murii = mur_iso, murij = 0	1	1
Electrical conductivity	sigma_iso ; sigmaii = sigma_iso, sigmaij = 0	5.998e7[S/m]	S/m
Relative permittivity	epsilonr_iso ; epsilonrii = epsilonr_iso, epsilonrij = 0	1	1

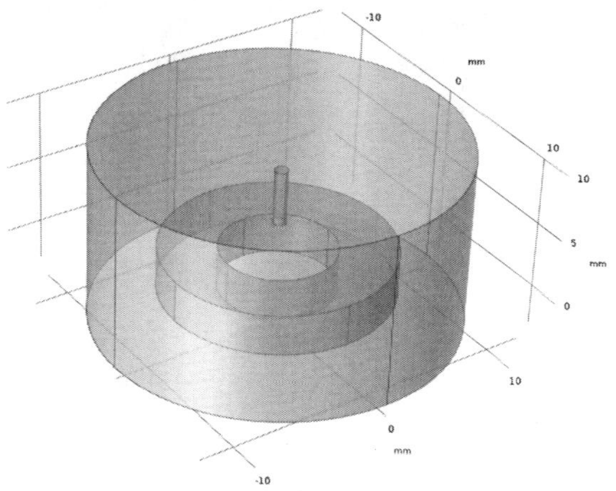

Figure 5.103 Material allocation for the tool as anode and coil

The boundary conditions for the first module, Magnetic Fields were set as surface current density on the coil, as it is presented in Figure 5.105.

Running the model for the first module, the magnetic field distribution was obtained within the column as it is presented in Figure 5.106.

The boundary conditions for the second module, Charged Particle Tracing consisted in Magnetic Force, generated by the magnetic flux within the column (Figure 5.107), determined

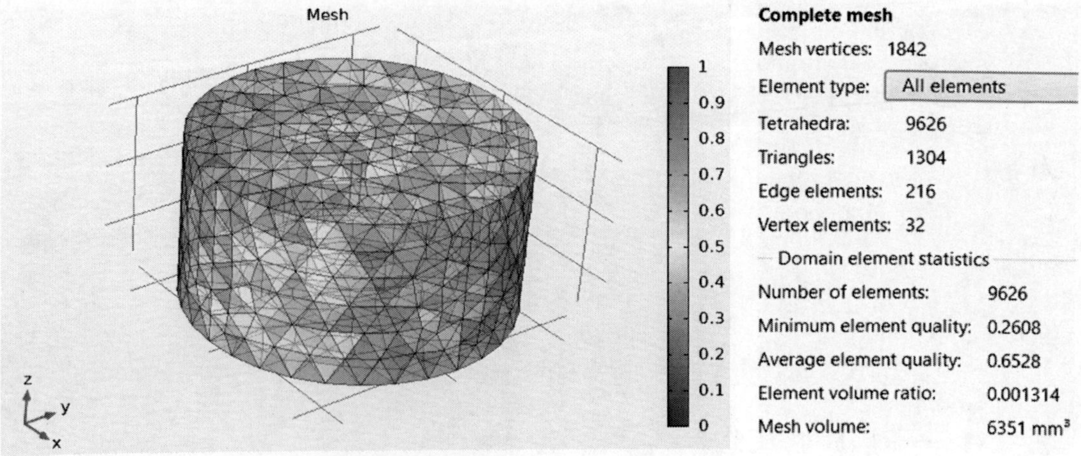

Figure 5.104 The mesh and its quality and statistics

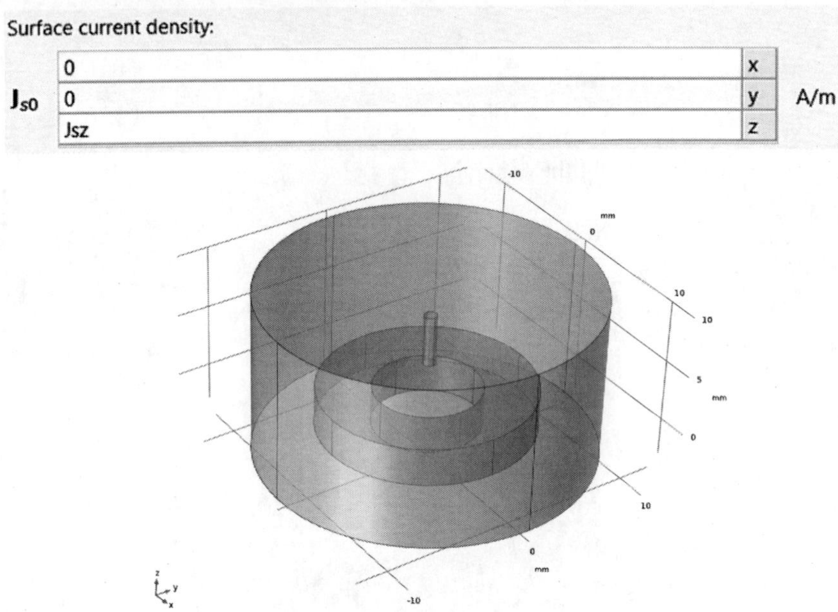

Figure 5.105 Surface current density on the coil

previously by running the first module and then inlet 1, respective the release of particles, copper ions at the end of the anode (Figure 5.108).

The particle entry area and study settings are both specified in this case. The settings from Figure 5.108 indicate uniform particle distribution, number of particles released from surface 13 (end of the tool as anode) and initial velocity.

As seen in Figure 5.109, after running the program, the ion trajectory is not deflected significantly, thus creating conditions for small area deposition of materials on the surface of the workpiece.

For the determination of the ion beam trajectory (deviation) in the working space, Poincaré maps were visualized based on the value of the current surface density (Jsz). Figures 5.110, 5.111, and 5.112 show the cut planes utilized for this purpose.

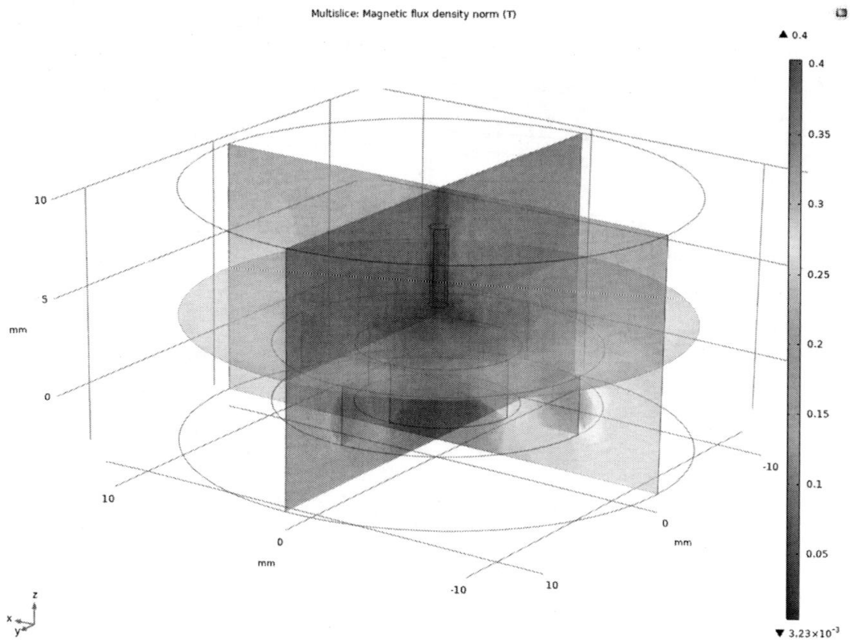

Figure 5.106 Magnetic flux density within the column

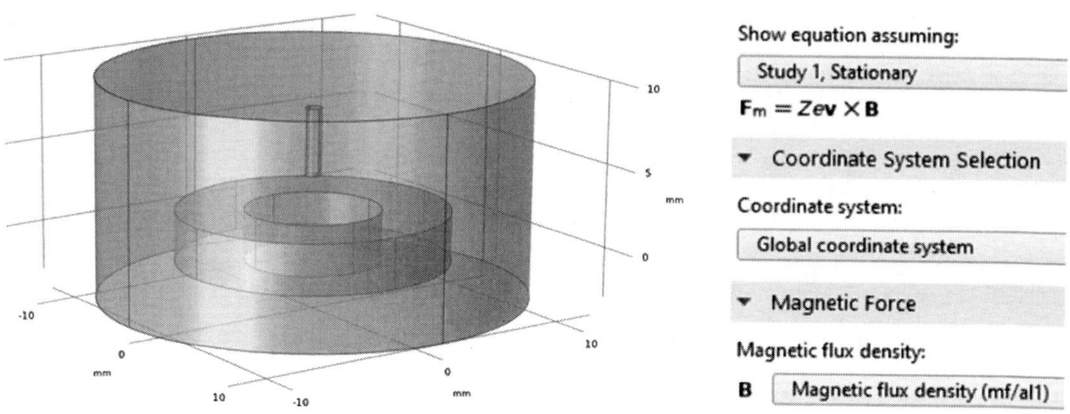

Figure 5.107 Magnetic force condition within the column

Discussion

The Jsz values were modified to reach the values of the magnetic flux as previously presented (Muralidharan et al., 2015), and the correspondence between these parameters is presented in Table 5.13, based on study 1 from Comsol Multiphysics™ modeling.

The Poincaré graphs from Figures 5.113 to 5.117, in relation to Jsz values, show an increase in deviation of the ion trajectory on the target surface as the current density increases. There is a corresponding color for each previous cut plane: $Z = 4$, red; $Z = 2$, black; $Z = -3$, blue.

It can be observed that when the surface current density is increased, a more irregular and larger distribution on the sample results in an optimal distribution at a mean magnetic flux

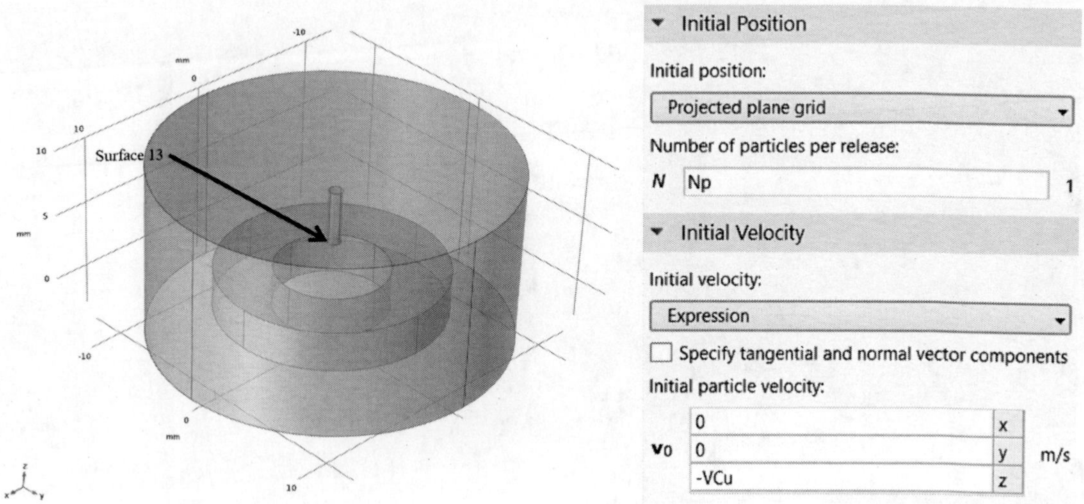

Figure 5.108 Particle inlet area and beam inlet settings

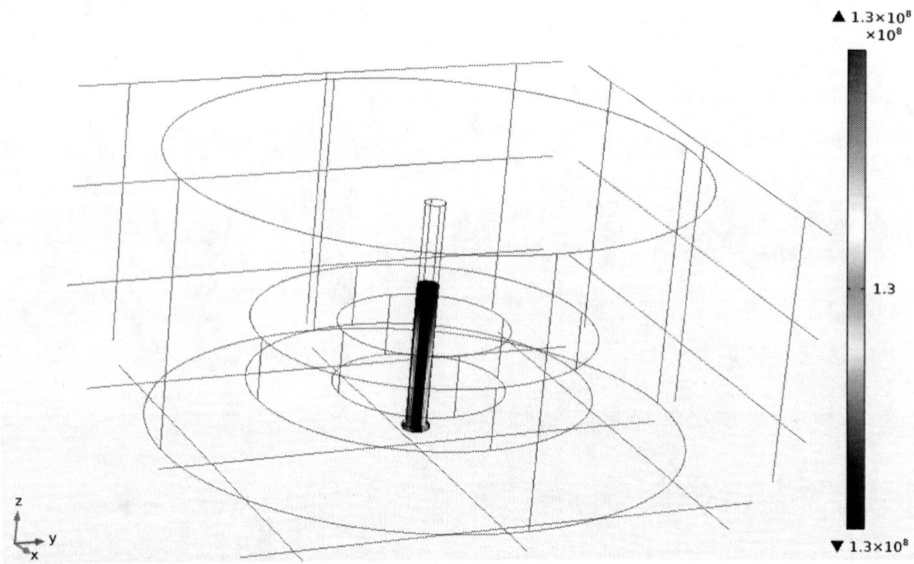

Figure 5.109 Particle trajectory simulation

value. This is a key parameter of the hybrid process of electrical discharge deposition aided by magnetic field.

Conclusion

The study of electrical discharge deposition is based on finite element modeling of the process that comprised two stages:

(1) the determination of the magnetic flux inside a column that represents the working space; (2) the determination of the charged particle trajectory, e.g., copper ions that are deposited on a steel part.

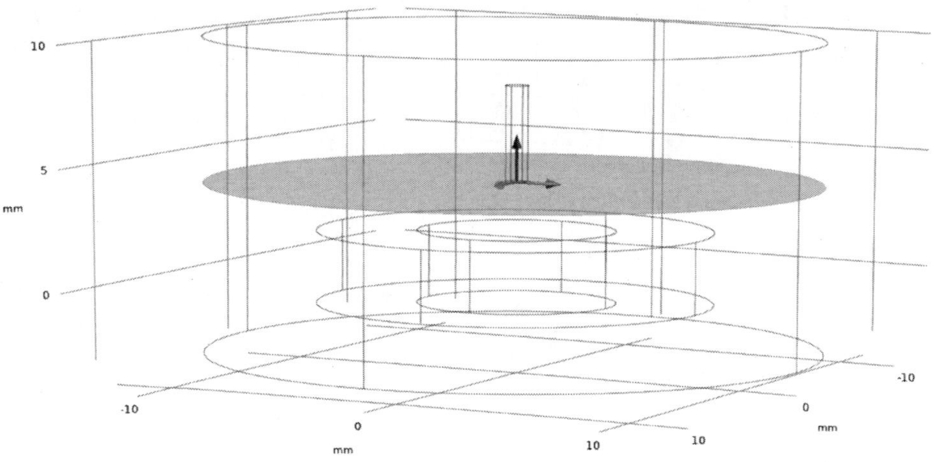

Figure 5.110 Cut plane 1 with coordinate $Z = 4$

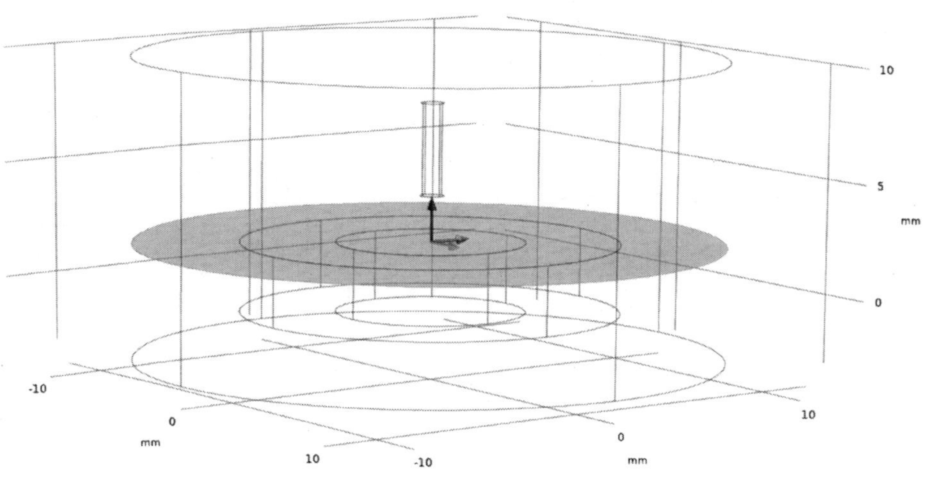

Figure 5.111 Cut plane 2 with coordinate $Z = 2$

Since the process of electrical discharge deposition is hybrid, the distribution of the material aided on the surface part is influenced by electrical parameters of the working mode, similar to electric discharge machining, as voltage, current, pulse and pause time, and polarity, but also by the current and the dimensions of the electromagnetic coil.

Further experiments are needed for a deeper understanding of the electrical discharge deposition mechanism and validation of the numerical simulation.

Modeling of Microfluidics

Microfluidic devices are of utmost importance nowadays, because they can be implemented on a lab-on-a-chip for counting different types of cells from the blood, like T lymphocytes. These types of leukocytes (white cells) are responsible for creating the human immunity system (Papenfuss et al., 2016). These lab-on-a-chips have the following main advantages: the analysis

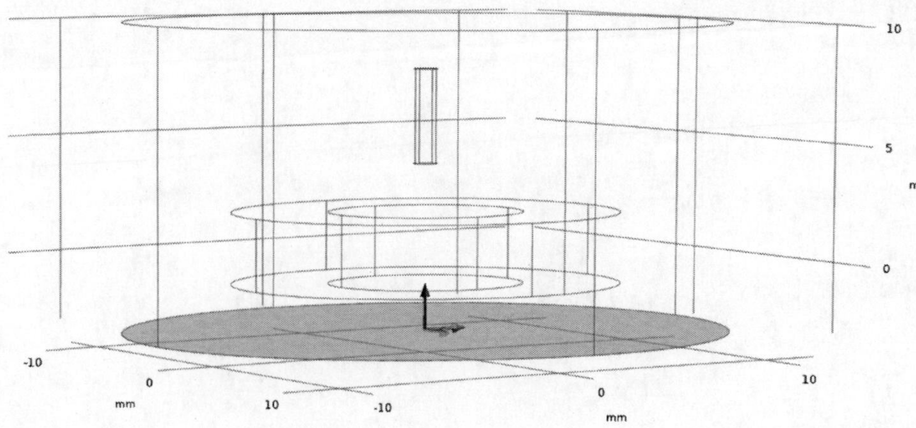

Figure 5.112 Cut plane 3 with coordinate $Z = -3$

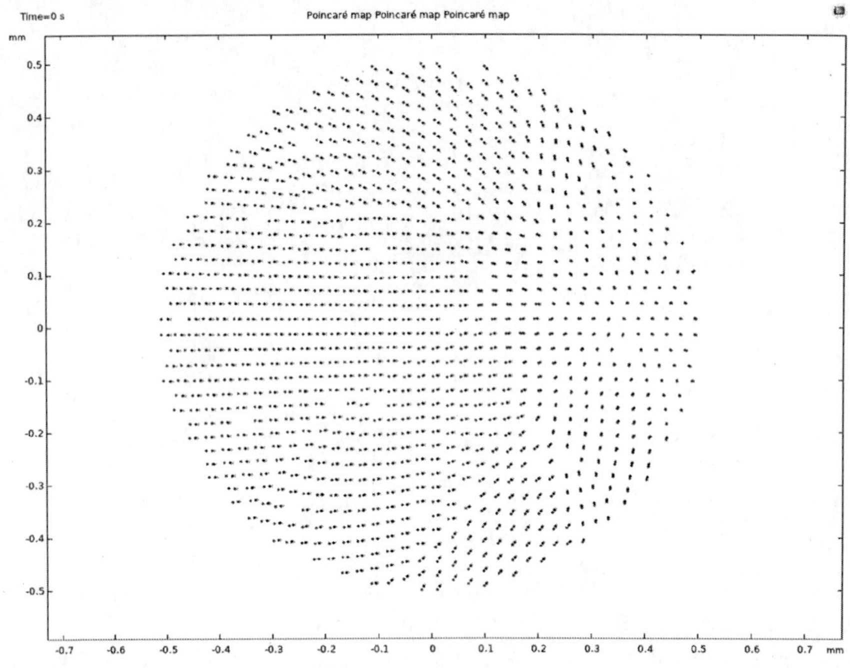

Figure 5.113 Poincaré map for Jsz = 60,000 A/m

Table 5.13 The correspondence between magnetic flux and surface current density

Parameters	Levels				
Magnetic flux (Tesla)	0.07	0.1	0.25	0.3	0.40
Surface current density (Jzs) (A/m)	60,000	90,000	220,000	270,000	360,000

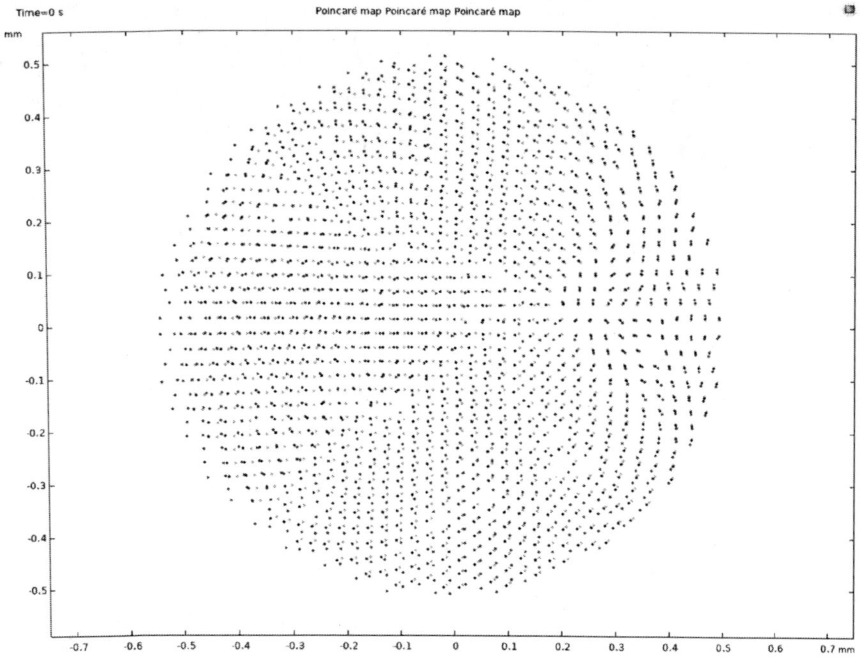

Figure 5.114 Poincaré map for Jsz = 90,000 A/m

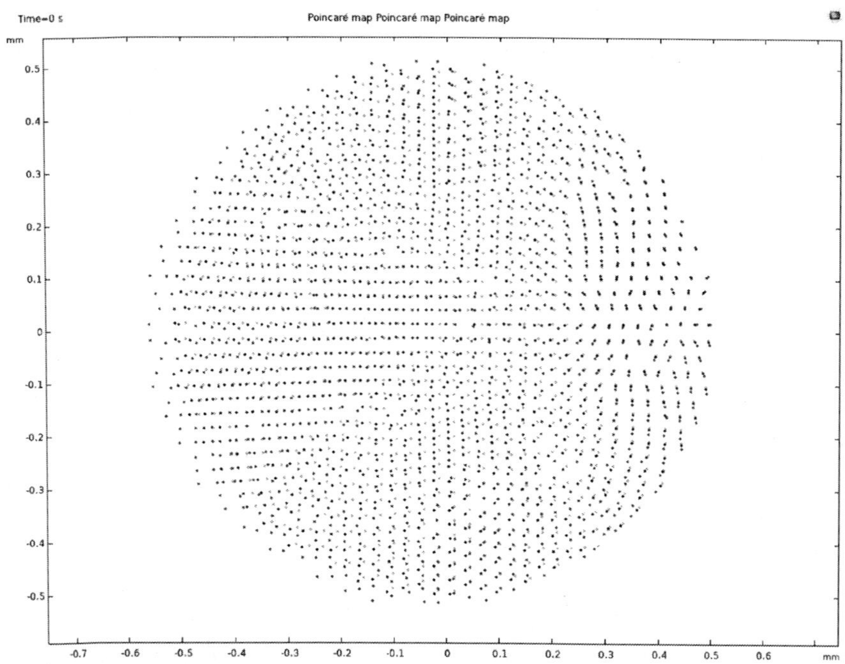

Figure 5.115 Poincaré map for Jsz = 220,000 A/m

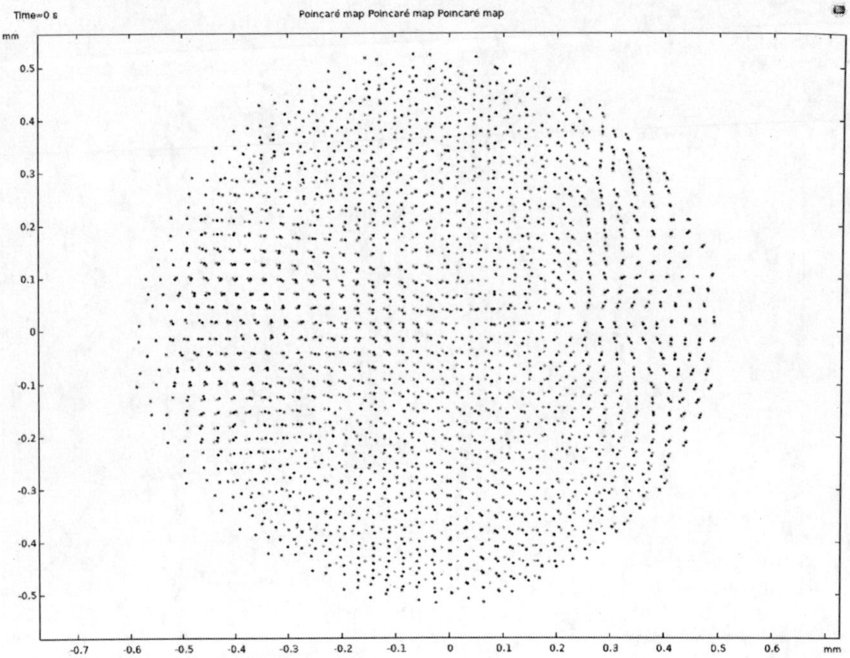

Figure 5.116 Poincaré map for Jsz = 270,000 A/m

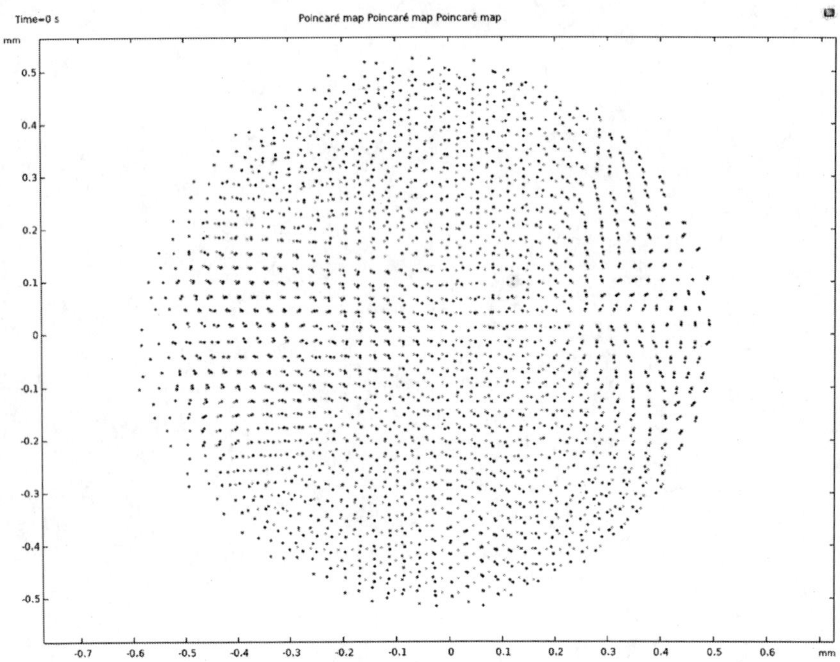

Figure 5.117 Poincaré map for Jsz = 360,000 A/m

results are given in a short time; analyses cannot be performed using other usual equipment and they can be reused. Other advantages include small amount of blood samples and reaction substances used, the possibility to connect to data acquisition systems, remote control, reliable results, and reduced manufacturing costs (Oh, 2020).

The microfluidic circuits, due to their small dimensions and their intricate shapes, can be achieved using only nonconventional technologies (concentrated energies technologies) like photochemical lithography, laser, and even 3D printing. This is based on the capacity of the non-conventional technologies to concentrate the energy, transferred from the tool, on very reduced space of the machined surface.

Some T cells like CD3+, CD4+, and CD8+ from the blood involved in lymphoid cancer and other diseases can be identified (Scherrenburg et al., 2008). In order to do this, an electrochemical sensor is used within the microfluidic circuits that counts the number of total lymphocytes that pass one by one through the counting channels, which have narrowed sections that correspond to specific cell dimensions. As each cell passes through the counting channel, a sensor collects the variation of electrical parameter, e.g., peak of voltage, and sends this to a data acquisition system. A capture chamber with pylons is used for retaining the cells that are not analyzed, whose dimensions are greater than the distance between the pylons.

Input Data for Modeling

The liquids that pass through the microfluidics devices are the blood samples, the substances of lysing the red cells (that facilitate the lymphocytes counting), and stopper lysing substances (that avoid the destruction of the lymphocytes that are to be counted). The substances mentioned above have the following flow proportions: 50 μl/min of blood, 1600 μl/min lysis reactive, and 265 μl/min solution for stopping the lysing. The lysing solution has the content: 0.05% (v/v) surfactant, 0.05% (v/v) acetic acid, and 0.15% (v/v) organic acid in deionized water. This was introduced in the first circuit of lysing of microfluidic device and required a reaction time lower than 3 s in order to eliminate only the red cells. The stopping lysing solution has three sodium salts in the following proportions: 3.13% (w/v) salt 1; 1.45 % (w/v) salt 2, and 0.6% (w/v) salt 3 in deionized water. This was introduced in the second circuit of lysing stopping and required a reaction time longer than 30 s for lymphocytes protection. The optimal ratio between the blood sample, lysing solution, and stopping lysing solution is 1:12:5.3. This ratio was kept using a micropump, for any volumes of the probes used.

Numerical Simulation

Comsol Multiphysics™ was used with the following succession: Fluid Flow, Single Phase Flow, and Laminar Flow. The water is considered as a single phase, because it represents more than 99% in the mix of blood sample and the aqueous reactive solutions. Laminar flow was also adopted after the calculation of Reynolds number (Janna, 2020):

$$R_e = \frac{\rho u D}{\mu} \tag{5}$$

where ρ—water density (kg/m³); u—average velocity (m/s) of the flow within the device's circuits; D—inside transversal dimension of the circuit (m), $D = 3 \times \cdot 10^{-4}$ (see parameter a, Figure 5.41); μ—dynamic viscosity (Pa·s); $\mu = 8.9 \times \cdot 10^{-4}$. The obtained values were 1.68 for circuit 1, and 0.79 for circuit 2. These are far lower than 2100, the limit between laminar and turbulent flow (Janna, 2020).

The numerical simulation consisted of the following stages:

a) the parameterization of the model (Figure 5.118)
b) the design of model geometry using Comsol™ built-in tools (Figure 5.119)
   The geometry created, based mostly on spiral shapes of the fluid circuits, has the advantages that it is compact, efficiently uses the space of a circular silicon wafer, and most important it reduces the fluid pressure loss.
c) The boundary conditions are three inputs and one output, Figures 5.120–5.123.
d) The mesh quality and its characteristics are presented in Figure 5.124.

The velocity distribution within the microfluidic circuits is presented in Figure 5.125.

The pressure variation within the microfluidic circuits is presented in Figure 5.126.

Finally, the fluid's velocity is collected in points from the median flowing path, due to parabolic distribution of laminar flow, belonging to the first (lysing)—Figure 5.127, and second (stopping of lysing) circuit—Figure 5.128.

▾ Parameters

Name	Expression	Value	Description
vs	0.001	0.001	input velocity of blood [m/s]
vl	vs*12	0.012	input velocity of lysing [m/s]
vsl	vs*5.3	0.0053	input velocity of lysing stopping [m/s]
pout	101325	1.0133E5	outpressure [Pa]
lin	100	100	input length of blood and lysing
a	300	300	channel width
rrac	300	300	curve radius
l1	1000	1000	length of construction horizontal channel
l5	200	200	output length
b	15	15	width of counting channel
b2	12	12	width of counting channel 2
l7	30	30	length of couting channel
l1cap	4002	4002	length of capture chamber
l2cap	3052	3052	width of capture chamber
rp	178/2	89	pylon radius
distp	12	12	distance between pylons

Figure 5.118 Parameterization of the model

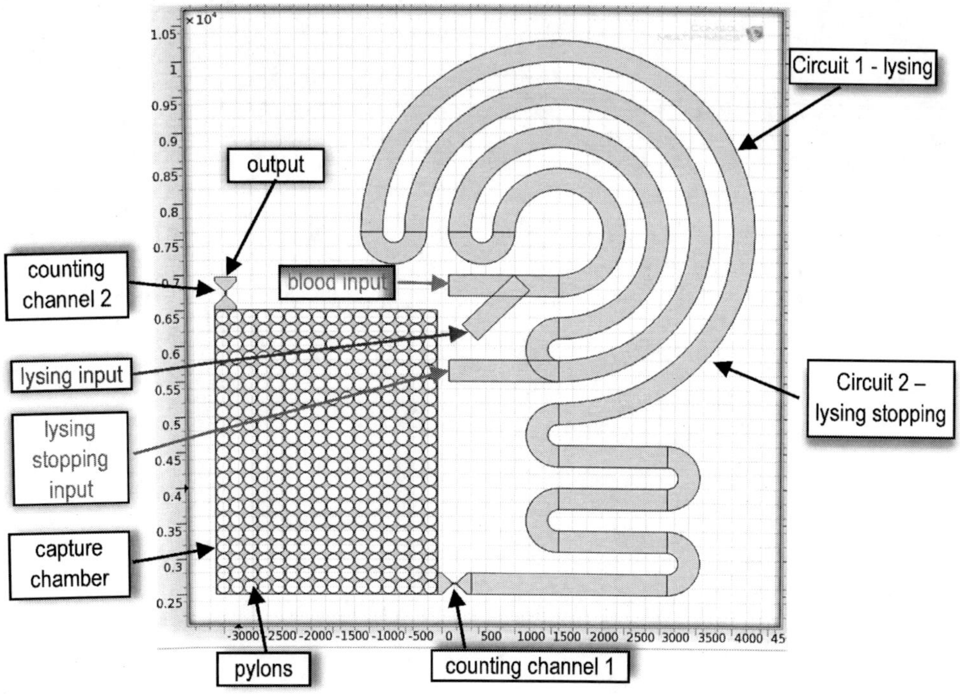

Figure 5.119 Geometry of the model

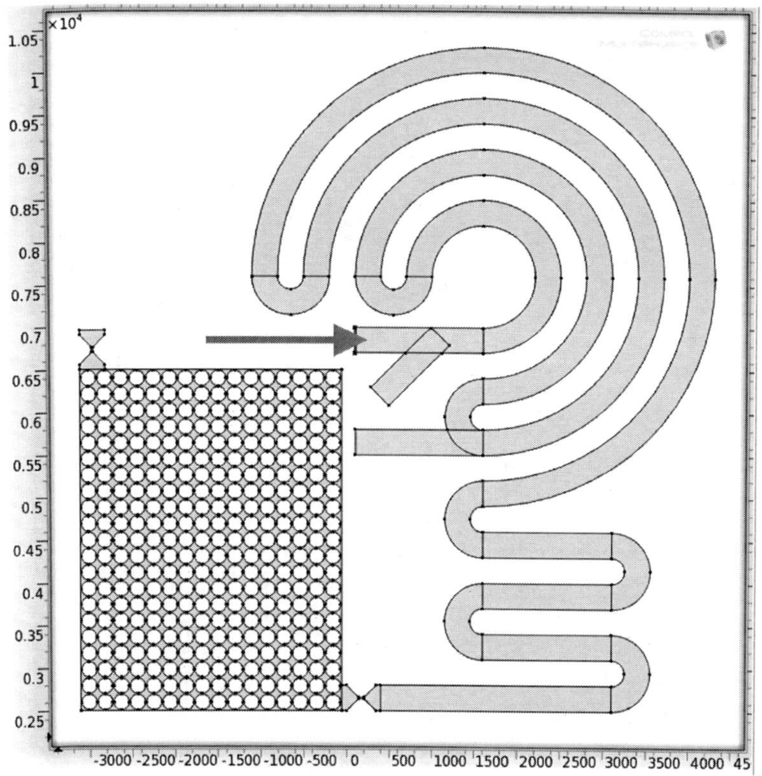

Figure 5.120 Input of the blood sample as velocity, *vs*

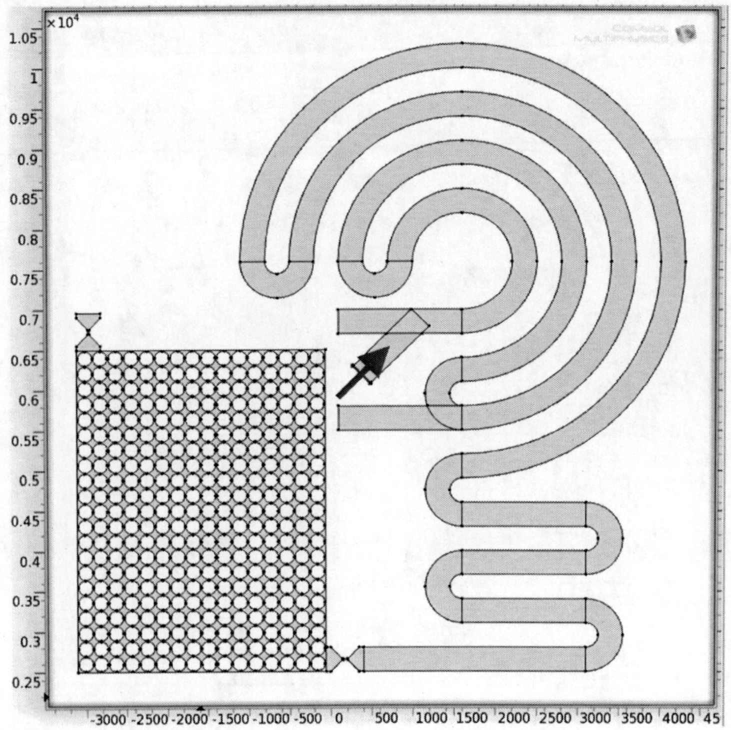

Figure 5.121 Input of the lysing liquid as velocity, *vl*

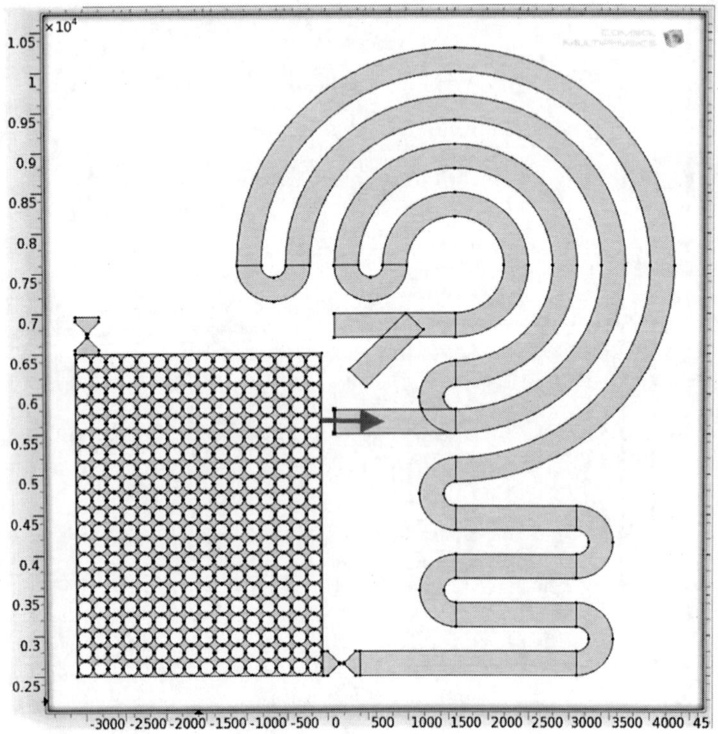

Figure 5.122 Input of the lysing stopping liquid as velocity, *vsl*

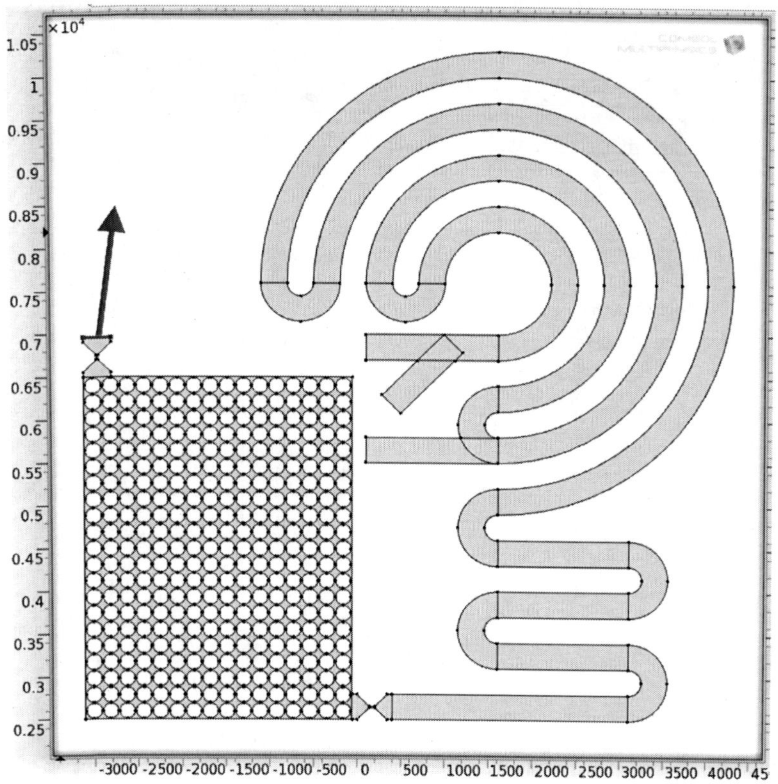

Figure 5.123 Output of the fluids as atmospheric pressure, *pout*

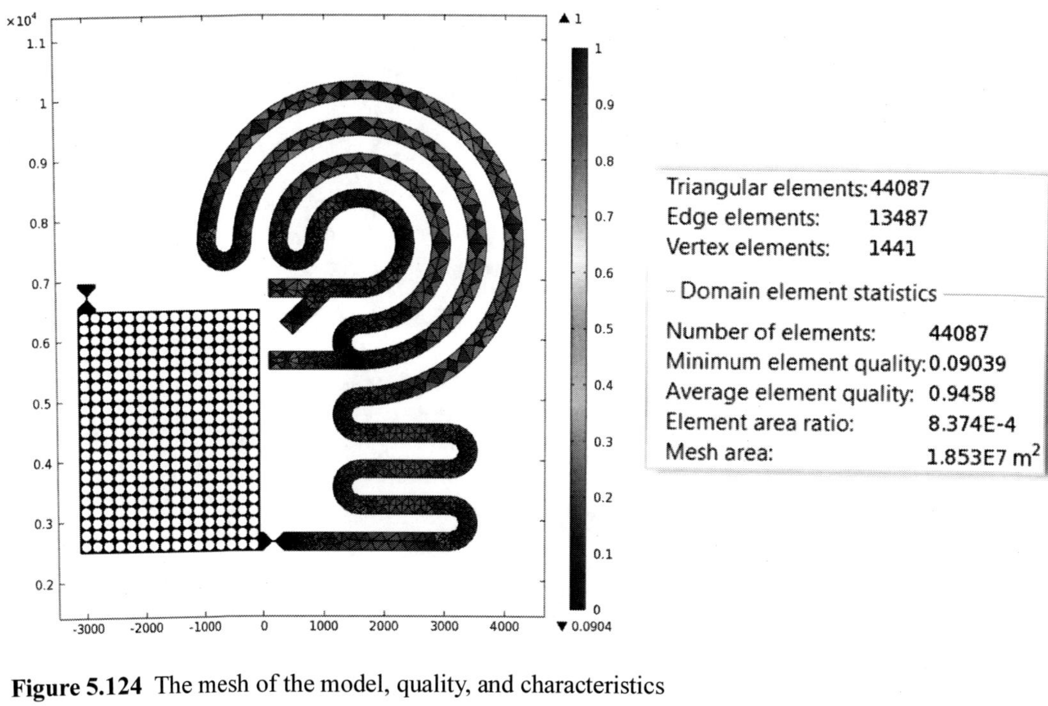

Figure 5.124 The mesh of the model, quality, and characteristics

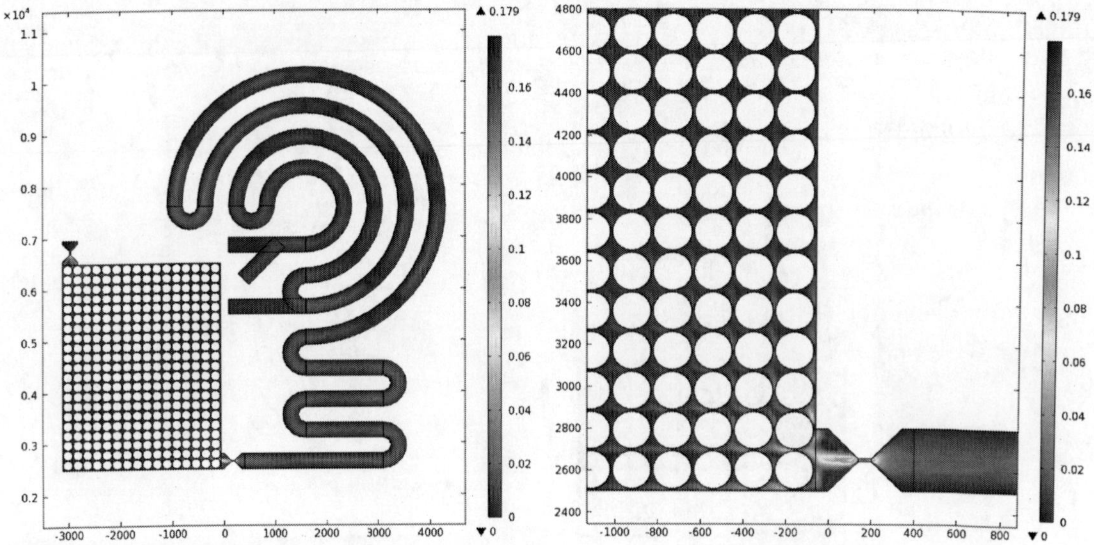

Figure 5.125 The fluids velocity distribution (m/s); left, general view, right, detail

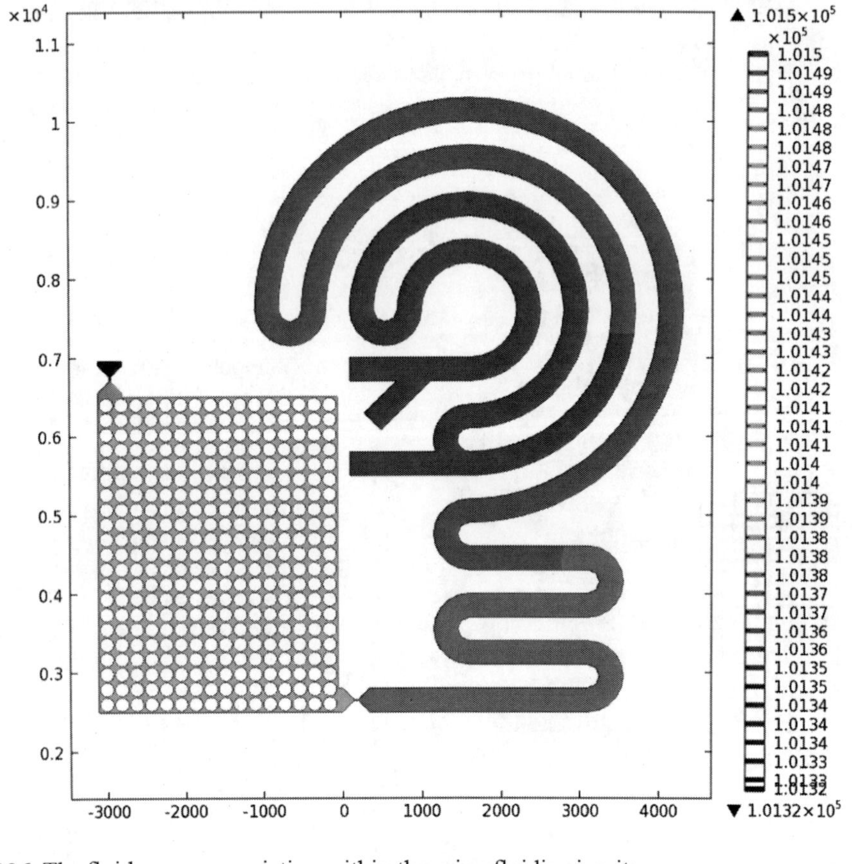

Figure 5.126 The fluid pressure variation within the microfluidic circuits

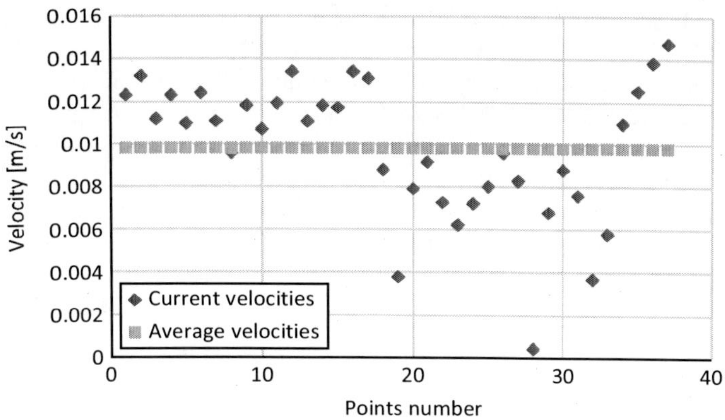

Figure 5.127 The flowing velocities within the lysing circuit (1)

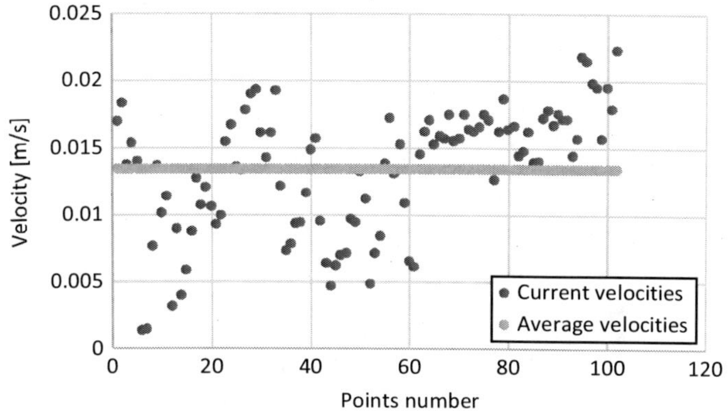

Figure 5.128 The flowing velocities within the stopping lysing circuit (2)

Conclusion

Based on the average velocities, the time is calculated in which the blood mixed with the lysing solution flows within the first circuit ($t_1 = 2.8$ s). The time needed for the liquids combined with the solution for stopping the lysing to go through the second circuit is also calculated ($t_2 = 46.2$). The values meet the requirements for reaction, lower than 3 s for circuit 1, and greater than 30 s for circuit 2. The results of the numerical simulation of microfluidics behavior were confirmed by the experimental data obtained by testing the microfluidic device.

References

Baron S, Ahearne E, Connolly P, Keaveney S, Byrne G. An assessment of medical grade cobalt chromium alloy ASTM F1537 as a "Difficult-to-Cut (DTC)" material. *Proceedings of the Machine Tool Technologies Research Foundation Annual Meeting*, San Francisco June 30–July 2, 2015.

Buser D, Sennerby L, De Bruyn H. Modern implant dentistry based on osseointegration: 50 years of progress, current trends and open questions. *Periodontology 2000* (2017) 73: 7–21.

Chatti S, Laperriere L, Reinhart G, Tolias T. *CIRP Encyclopedia of Production Engineering.* Berlin: Springer, 2019.

Diego AV, Rogelio LH, Fernando JV, Gustavo MF. Modeling and vibration mode analysis of a ball screw drive. *International Journal of Advanced Manufacturing Technology* (2012) 58: 257–265.

Gherbesi E, Natalini G. The ultimaster coronary stent system: 5-year worldwide experience. *Future Cardiology* (2020) 16: 251–261.

Ghiculescu D. *Nonconventional Machining.* Bucharest: Printech, 2004.

Ghiculescu D. *Computer Aided Engineering and Manufacturing in Nonconventional Technologies Field.* Bucharest: Printech, 2013.

Ghiculescu D, Marinescu I, Jitianu G. On precision improvement by ultrasonics-aided electrodischarge machining. *Estonian Journal of Engineering* (2009) 15: 24–33.

Jahan MP et al. *Electrical Discharge Machining (EDM): Types, Technologies and Applications.* New York: Nova Science, 2015.

Janna WS. *Introduction to Fluid Mechanics*, Sixth Edition (6th ed.). CRC Press. Boca Raton, 2020. https://doi.org/10.1201/9780429324536.

Jitendra K, Janakarajan R, Dhamodaran S, Kanmani S. Dry micro-electric discharge deposition of copper on die steel: Effect of pulse on-time. *International Workshop on Micro Factories* (2010) 7: 1–4.

Leyens C, Peters M. *Titanium and Titanium Alloys: Fundamentals and Applications.* Weinheim: WILEY-VCH, 2003.

Liu B, Liu Y. Powder metallurgy titanium aluminide alloys. *Science, Technology and Applications* (2015) 27: 515–531.

Liu G, Wang X, Zhou X., Zhang L, Mi J, Shan Z, Huang B, Chen Z, Chen Z. Modulating the cobalt dose range to manipulate multisystem cooperation in bone environment: A strategy to resolve the controversies about cobalt use for orthopedic applications. *Theranostics* (2020) 10: 1074–1089.

Liu Q, Zhang Q, Zhang M, Zhang J. Review of size effects in micro electrical discharge machining. *Precision Engineering* (2016) 44: 29–44.

Louwerens JK, Hockers N, Achten G, Sierevelt IN, Nolte PA, Van Hove RP. No clinical difference between TiN-coated versus uncoated cementless CoCrMo mobile-bearing total knee arthroplasty; 10-year follow-up of a randomized controlled trial. *Knee Surgery Sports Traumatology Arthroscopy* (2020) 29: 1–7.

Makenzi M, Ikua B. A review of flushing techniques used in electrical discharge machining. *Proceedings of the 2012 Mechanical Engineering Conference on Sustainable Research and Innovation* (2012) 4: 162–165.

Marinescu NI, Ghiculescu D. *Electrochemical Machining and Related Processes.* Bucharest: Printech, 2005.

Marinescu NI, Ghiculescu D, et al. *Technological Processes with Beams, Oscillations and Jets.* Bucharest: Printech, 2019.

Mehdi H, Ahmed ADS, Mohd S. Micro-electrode fabrication processes for micro-EDM drilling and milling: A state-of-the-art review. *The International Journal of Advanced Manufacturing Technology* (2017) 91: 1023–1056.

Mori M et al. Microstructures and mechanical properties of biomedical Co-29Cr-6Mo-0.14N alloys processed by hot rolling. *Metallurgical and Materials Transactions* (2012) 43: 3108–3119.

Murlidharan B, Chelladurai H, Janakarajan R, Experimental investigation on electro-discharge deposition process. *The International Journal of Advanced Manufacturing Technology* (2015) 76: 69–82.

Oh KW. Microfluidic devices for biomedical applications: Biomedical microfluidic devices 2019. *Micromachines* (2020) 11: 370.

Orang VA. Friction-Induced Vibration in Lead Screw Systems: Mathematical Modeling and Experimental Studies. *Journal of Vibration and Acoustics.* (2009) 131(2).

Oriental Motor General Catalogue. *Stepping motor and Drive Package DC Input*, C-121.

Papenfuss T, Bolon B. Lymphocytes. In: Vohr HW (eds) *Encyclopedia of Immunotoxicology.* Springer, Berlin, Heidelberg, 2016. https://doi.org/10.1007/978-3-642-54596-2_925.

Paulo Davim J. *Nontraditional Machining Processes: Research Advances.* London: Springer, 2013.

Rades M. *Mechanical Vibrations (in Romanian).* Bucharest: Printech, 2008.

Radovanovic M, Dasic P. Research on surface roughness by laser cut. *The Annals of University "Dunarea de Jos" of Galati* (2006) Fascicle VIII, Tribology.

Scherrenburg J, Piriou ERWAN, Nanlohy NM, Van Baarle D. Detailed analysis of Epstein-Barr virus-specific CD4+ and CD8+ T cell responses during infectious mononucleosis. *Clinical & Experimental Immunology* (2008) 153(2): 231–239.

Senthil Kumar V, Dr. Jayaprakash G. State of art of laser cutting process. *International Journal for Modern Trends in Science and Technology* (2017) 3(4): 76–80.

SKF Group. *A Selection Guide*. PUB MT/P1 14489 EN.

Szkliniarz W, Szkliniarz A. Microstructure and properties of TiAl-based alloys melted in graphite crucible. *Metals* (2021) 11: 669–686.

Vaicelyte A, Janssen C, Le Borgne M, Grosgogeat B. Cobalt–chromium dental alloys: Metal exposures, toxicological risks, CMR classification, and EU regulatory framework. *Crystals* (2020) 10: 1151.

Wang ZL, Cui J, Jin B. Study on Micro Electrical Discharge Deposition in Air, *2007 International Conference on Mechatronics and Automation*, Harbin, China, 63-68 doi: 10.1109/ICMA.2007.4303517.

Zhou Y, Peng F, Chen J. Torsion Vibration Analysis of Lead-screw Feed Drives with Changeable Table Position and Work-piece Mass, 2007 *International Conference on Mechatronics and Automation*, Harbin, China, 2007, pp. 2194-2199, doi: 10.1109/ICMA.2007.4303892.

Zilong P, Guan X. Micro electrical discharge machining deposition in air for fabrication of micro spiral structures. *Chinese Journal of Mechanical Engineering* (2010) 23(2): 154–160.

Chapter 6

Evaluation of Gear Flanks Using Gear Topography Data

Nikolaos Tapoglou,[1,*] Anastasios Tzotzis,[2] Panagiotis Kyratsis,[2] Chara Efstathiou[3]

[1] Department of Industrial Engineering and Management, International Hellenic University, Sindos, Thessaloniki, Greece

[2] Department of Product and Systems Design Engineering, University of Western Macedonia, Kila, Kozani, Greece

[3] Department of Production Engineering and Management, Technical University of Crete, Greece

Abstract

High-quality transmission components are essential in a series of engineering systems. Involute gears are at the core of every transmission system and can be manufactured with a large variety of manufacturing methods. Each gear manufacturing method offers different productivity and quality outcomes. The topography of gears manufactured with different methods is considerably different and thus leads to deviations in the performance during operation. In an effort to understand the deviations of machined gears, the research presented focuses on the development of an analysis and visualisation platform for gear flank topographical data. The platform is tested with the use of data from a series of gear manufacturing processes.

Keywords: gear metrology, power skiving, gear flank, CAD

Introduction

Over the last decade, the requirements for gear transmissions are becoming more stringent, owing to the drive towards more efficient and sustainable engineering systems. The aerospace and automotive sectors are the primary drivers behind this trend. On one side the move towards higher bypass ratios and on the other the growth of electric propulsion systems lead to increased requirements from transmission systems. Transmission systems have to deliver increased torque with reduced weight and in a smaller environment. The quality of gear forms is a critical product of every gear manufacturing process. In order to achieve high-efficiency transmission, low noise output and long component life, the deviations of involute gears must be thoroughly investigated. Different gear manufacturing processes, due to the kinematic chain

* Corresponding Author Email: ntapoglou@iem.ihu.gr

they are based on, result in considerably different flank topographies. In addition to this, gear manufacturing processes are traditionally performed with complex kinematics that is influenced by a series of parameters, all of which have a different effect on the topography of the gear.

The quality characteristics of machined gears have been the focus of many researchers due to their criticality in the manufacturing process chain. Researchers have investigated the effect of the cutting process on the gear topography, the way the gears are measured as well as the standards followed.

Goch (2003) presented the most comprehensive research on the field of gear metrology. In the state-of-the-art review, he performed the key gear measurement methods analyses along with the methodology for extracting key information from the raw measurement data. The review focused on both cylindrical and bevel gear geometries.

Michalski (2009) presented an analysis of gear topographies of gears manufactured through gear hobbing and shaping. In his research, the tooth flanks were presented using greyscale maps so that the three-dimensional state of the machined surface can be visualised. In the greyscale maps, key features of bot processes were revealed.

Moru and Borro (2020) presented a machine vision algorithm for the inspection of gear components, while Yin et al. (2022) focused on the influence of errors during the establishment of the coordinate system in the tooth profile deviations.

Digital twins and simulation models have been increasingly used in the design and optimisation of manufacturing processes. This is also the case for a series of gear manufacturing processes (Bouzakis et al. 2008, Tapoglou and Antoniadis 2012, Tapoglou 2019, 2021, Tapoglou and Kyratsis 2022). Accurate analysis and visualisation post-processing algorithms are required for the efficient use of such digital twin systems in order to maximise the use of the information available.

Extending the research on digital twins, this research introduces a novel architecture for the analysis and visualisation of gear manufacturing data that derive from CAD-based simulation models. A series of scenarios are examined in a series of gear manufacturing methods.

The remainder of the chapter is structured as follows; the next section presents the architecture of the developed platform as well as the structure and results of the algorithms implemented, followed by the implementation of the simulation platform. The final section introduces the concluding remarks and future work.

System Architecture

In order to support the post-processing of gear manufacturing simulation and inspection, the platform developed works in tandem with a CAD computational kernel in order to leverage the advanced accuracy and representation capabilities of such platforms. The architecture of the system is presented in Figure 6.1.

The platform receives as input a gear geometry in the form of a CAD-compatible file along with a series of values that provide the main geometrical characteristics of the gear to be inspected. The data are processed by the geometrical kernel of the platform, and the correct positioning and alignment of the gear are set in order to allow for the correct virtual inspection of the gear. Parallel to this step a virtual prototype of the gear geometry is generated. This

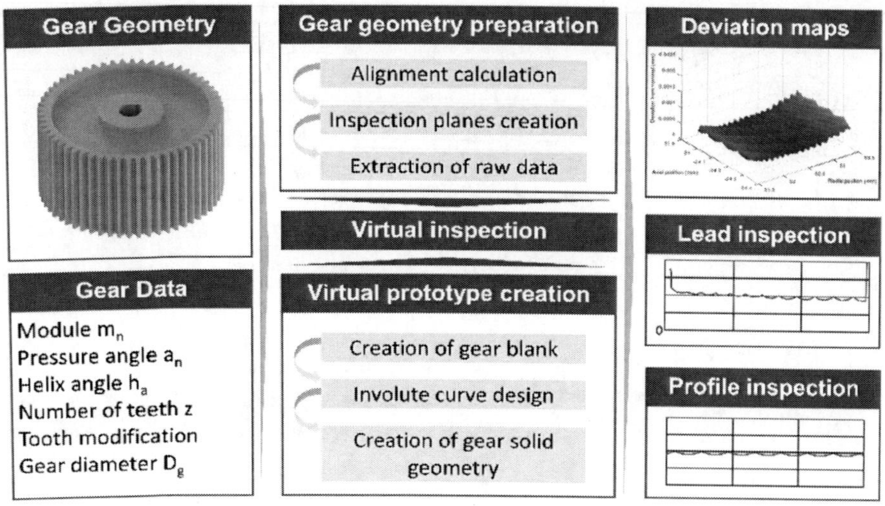

Figure 6.1 Inspection system architecture

would serve as a benchmark for the steps that follow. Next, the deviations of the test gear in relation to the prototype are extracted in the CAD environment. This allows for the maximum possible accuracy. A series of cross-sections on the profile and lead direction allow a complete understanding of the deviations of the two gears tested. The data are used to create deviation maps that highlight the deviations, their form as well as characteristics of the process such as feed marks.

The user can also execute a series of metrology investigations in order to calculate established gear tolerance values. The profile and lead directions can be investigated, and slope, form and total deviations can be calculated. Figure 6.2 presents the key deviation metrics for cylindrical involute gears based on ISO 1328 (2013). On the bottom side of the figure, the profile deviations can be observed while on the top the lead direction deviations are presented.

Case Studies

The digital metrology platform was used to investigate the deviations observed in two case studies. The first case study relates to a gear produced through the gear planning method through a CAD bases simulation approach (Tapoglou and Kyratsis 2022). The results of the examination of the gear through the platform are presented in Figure 6.3. On the left side, the 3D deviation map is presented. As can be observed, the gear is characterised by feed marks that run across the length of the gear flanks. On the right side of the figure, the inspection results for the profile and lead direction are presented. As it is expected, the lead inspection provides a fixed deviation from the nominal value, owing to the kinematics of the process. On the other hand, the profile inspection shows the peaks and valleys of the feed marks that run across the length of the gear.

The second case study involved the emerging process of power skiving. Similar to the first case study the virtually machined gear was acquired through a CAD-based simulation model. Figure 6.4 presents the deviation of the case study both in a 3D deviation map and in the form of profile and lead inspections. As can be observed, the marks on the surface of the flank are more complex when compared to the gear planning case. There is a series of feed marks that are

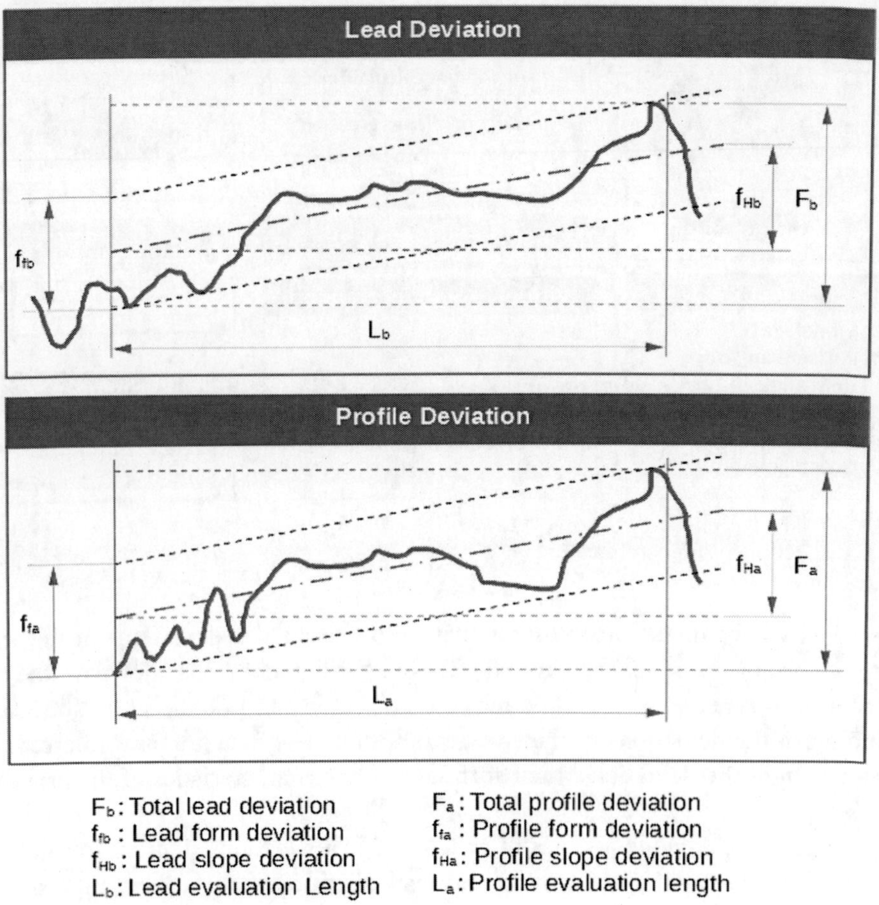

Figure 6.2 Gear deviation metrics

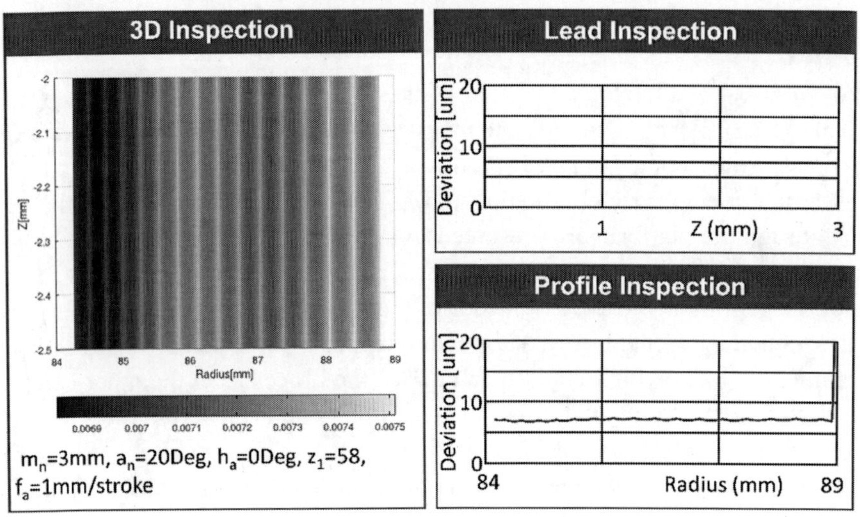

Figure 6.3 Inspection results of gear planning simulation

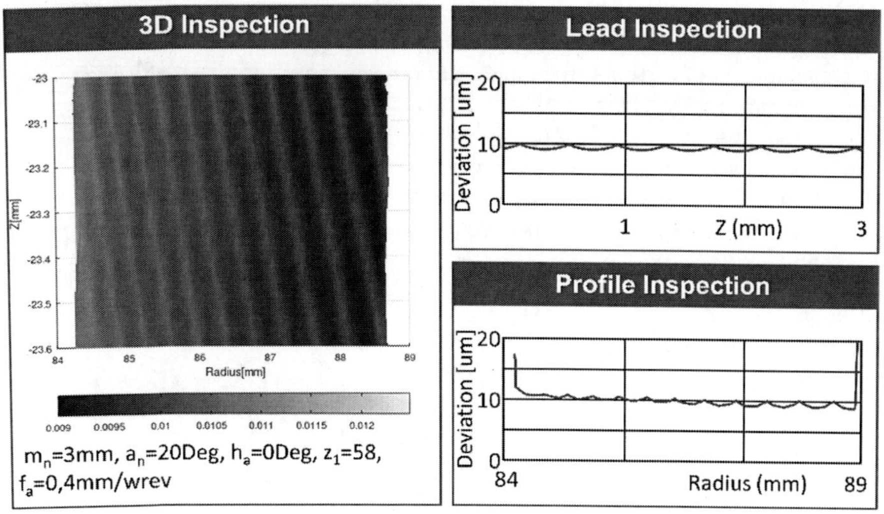

Figure 6.4 Inspection results of power skiving simulation

inclined in relation to the axis of the gear. This is typical for the power skiving manufacturing process. Looking at the lead inspection data the data reveals a periodic form of peaks and valleys that corresponds to the feed per revolution used.

Conclusion

The work presented focuses on the development of an architecture for the analysis and presentation of inspection results from CAD-based inspection data. The presented work helps in the understanding of manufacturing processes and the deviations observed due to the kinematics and the process parameters of the processes. The work is best positioned to support gear manufacturers in the optimisation of manufacturing processes. Further development of the platform would allow inspection data from metrology platforms to be integrated with the simulated measurements to allow for a holistic view of gear manufacturing processes.

References

Bouzakis, K.D., Lili, E., Michailidis, N., Friderikos, O. Manufacturing of cylindrical gears by generating cutting processes: A critical synthesis of analysis methods. *CIRP Annals — Manufacturing Technology* (2008) 57: 676–696.

Goch, G. Gear metrology. *CIRP Annals* (2003) 52(2): 659–695.

ISO. Cylindrical gears—ISO system of flank tolerance classification—part 1: Definitions and allowable values of deviations relevant to flanks of gear teeth: ISO 1328–1: 2013. Geneva: ISO, 2013.

Michalski, J. Surface topography of the cylindrical gear tooth flanks after machining. *The International Journal of Advanced Manufacturing Technology* (2009) 43(5): 513–528.

Moru, D.K., Borro, D. A machine vision algorithm for quality control inspection of gears. *The International Journal of Advanced Manufacturing Technology* (2020) 106(1): 105–123.

Tapoglou, N. Calculation of chip and gear geometry in power skiving using a CAD based simulation. *Journal of Advanced Manufacturing Technology* (2019) 100: 1779–1785.

Tapoglou, N. Development of cutting force model and process maps for power skiving using CAD-based modelling. *Machines* (2021) 9(5): 95.

Tapoglou, N., Antoniadis, A. CAD-based calculation of cutting force components in gear hobbing. *Journal of Manufacturing Science and Engineering* (2012) 134: 031009–1-8.

Tapoglou, N., Kyratsis, P. Investigation of gear profile deviations in gear planning process through CAD-based simulation. In *Recent Advances in Manufacturing Modelling and Optimization* (2022). Springer, Singapore, 89–94.

Yin, P., Han, F., Wang, J., Lu, C., Du, H. Influence of coordinate system establishment error on tooth profile deviation measured by standard polar coordinate method on gear measuring center. *Measurement* (2022) 187: 110344.

Chapter 7

CAD Plotters

Nemanja Kašiković*, Saša Petrović, Gojko Vladić, Gordana Bošnjaković, Željko Zeljković
Department of Graphic Engineering and Design, University of Novi Sad, Novi Sad, Serbia

Abstract

This book chapter will present the essential printing technologies for Computer-Aided Design (CAD) printing. In addition, the following material will give better insight into the necessary elements and basic principles of the mentioned printing processes, printing machine constructions with examples, and others. The focus of the chapter will be on the CAD plotters. This chapter compares CAD plotters based on different criteria, while emphasizing their best features as well as shortcomings. After a detailed comparison of the most popular and available CAD plotter technologies, it can be concluded that inkjet plotters are getting more and more advanced, thus becoming more popular than their toner-based counterparts in all areas.

Keywords: electrophotography, inkjet, plotters

Introduction

Printing techniques have two types of processes: conventional and digital. Conventional printing techniques, although vastly different from each other, have the 'printing plate'. A printing plate is a medium for ink transfer on the printing substrates. There are four main conventional printing techniques: Letterpress, Gravure, Planographic, and Screen Printing (Kipphan, 2001).

Digital printing techniques do not require a permanent master. They can be divided into a few main groups according to the physical or chemical principle they are based on (Kipphan, 2001): Electrophotography, Inkjet, Ionography, Magnetography, Thermography, Photography, X-graphy, and others.

Electrophotography and inkjet are the most popular printing techniques for CAD printing (a specialized printing process associated with CAD design).

This chapter will describe the basics of those printing processes and the printing equipment necessary for this production.

* Corresponding Author Email: knemanja@uns.ac.rs

Electrophotography

Electrophotography was invented in 1938 by Chester Carlson, an American physicist and inventor. Carlson submitted the invention in 1939 and obtained a patent in 1942 (Figure 7.1). Electrophotography involves five steps (Novaković and Kašiković, 2013; Kipphan, 2001; Kašiković et al., 2016; Majnarić, 2015):

1. Imaging
2. Inking
3. Toner transfer (printing)
4. Toner fixing
5. Cleaning (conditioning)

The core component in electrophotography is the photoconductor (Océ, 2006). OPC (organic photoconductor) is the most popular photoconductor type, although some printing systems can use arsenic triselenide (As_2Se_3) photoconductor or amorphic silicon.

In the first step, the surface of the photoreceptor is given a uniform electric charge using a corotron (positive electric charge), scorotron (negative electric charge), or dicotron (positive/negative electric charge) device. Light is then directed onto the surface of the charged photoreceptor, thereby selectively discharging it. LED arrays, spinning polygonal mirrors, DMD, and low viscosity toner (LVT) systems can be found in the electrophotographic process (Kipphan, 2001), and the final exposing result is a latent image made of areas with and without an electrostatic charge on the photoreceptor (Figure 7.2).

The inking step is the second step in electrophotography (Figure 7.3). In this stage, dry powder toner or liquid ink particles brought to the drum surface are either repelled or attracted to it, depending on the local surface charge and the toner's charge.

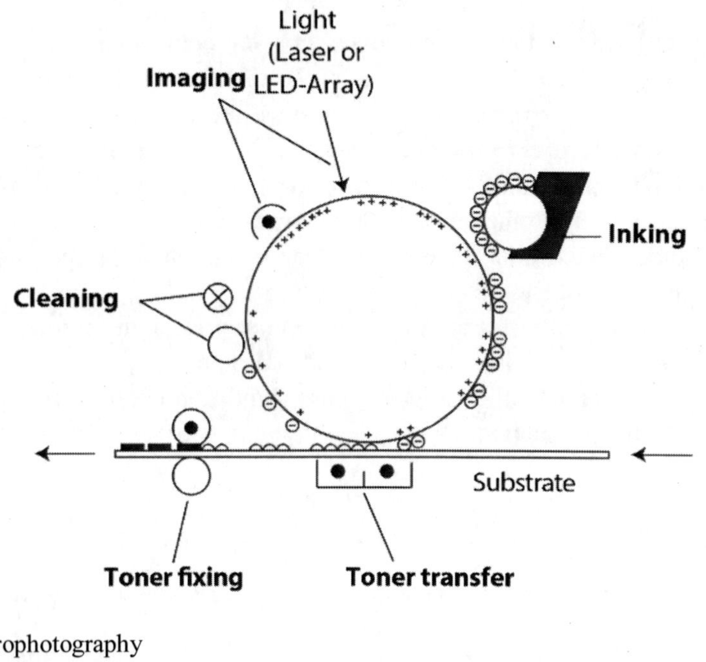

Figure 7.1 Electrophotography

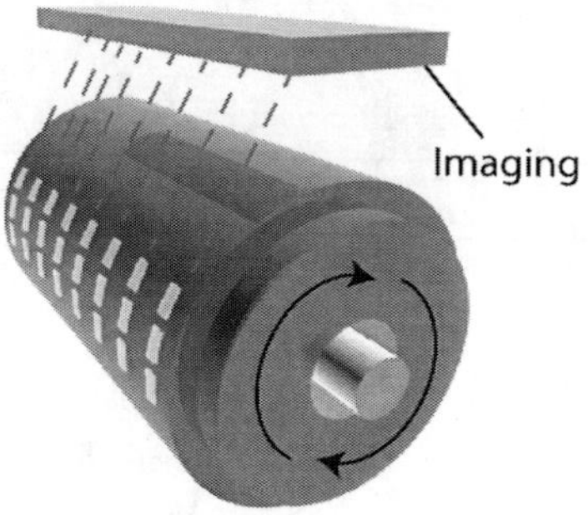

Figure 7.2 Imaging

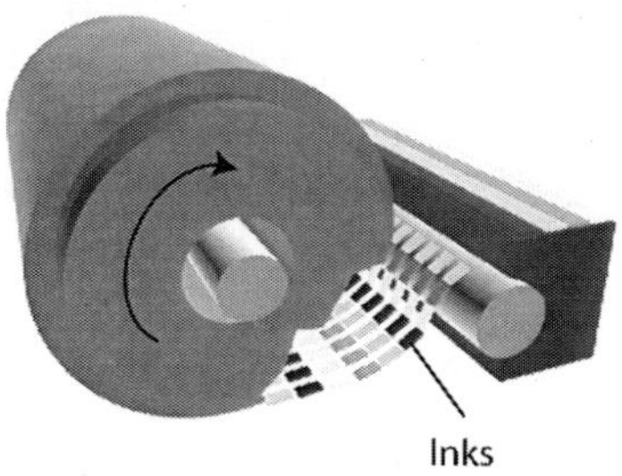

Figure 7.3 Inking

In electrophotography, it is nonessential to transfer the toner image from the photoreceptor surface to the media (Figure 7.4). The most common substrate, paper, is brought into contact with the drum at the transfer nib in this printing process. The clean transfer provides a second corona device (under the paper).

After transferring the toner to the substrate, it is necessary to fix a toner (Figure 7.5). The toner can be fused to the paper surface as it passes through two pressure rollers, one or both of which are heated. Radiant heat from a lamp or a high-intensity flash may also be used, although less common.

The last step in the electrophotography process is cleaning (Figure 7.6). In this stage, the photoreceptor drum surface is cleaned of remnants of toner. The cleaning process can be done with a brush, and any residual charge is removed with a lamp or electric discharge in preparation for a new imaging cycle.

Inkjet Printing

The most commonly used digital printing technology for CAD printing is inkjet. This process is a computer-to-print technology in which ink is sprayed from nozzles. That means no printing master is needed, and the imaging is done directly onto the substrate (Kipphan, 2001). That is why this is the shortest route for printing information on the substrate. Continuous inkjet (CIJ) and drop-on-demand inkjet are two main inkjet groups (Figure 7.7).

The main difference between these processes is that in drop-on-demand inkjet processes, drops of ink are only generated if the information to be printed requires them. Reversing it in the continuous inkjet, only a part of the continuously generated flow of tiny ink drops is directed

Figure 7.4 Toner transfer

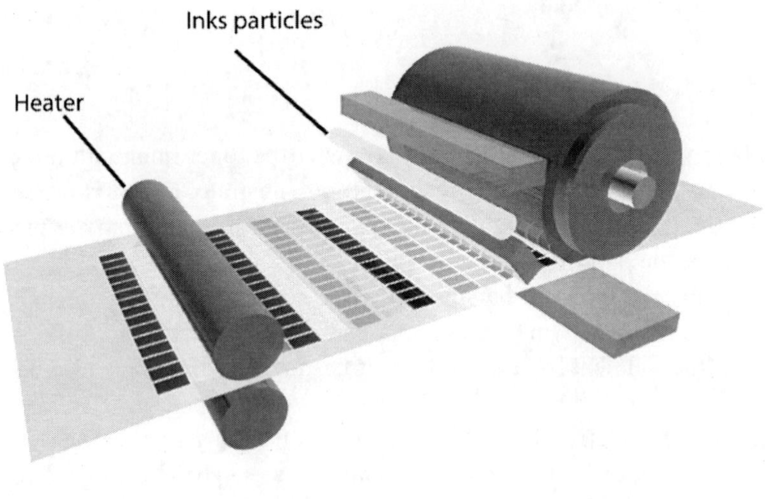

Figure 7.5 Toner fixing

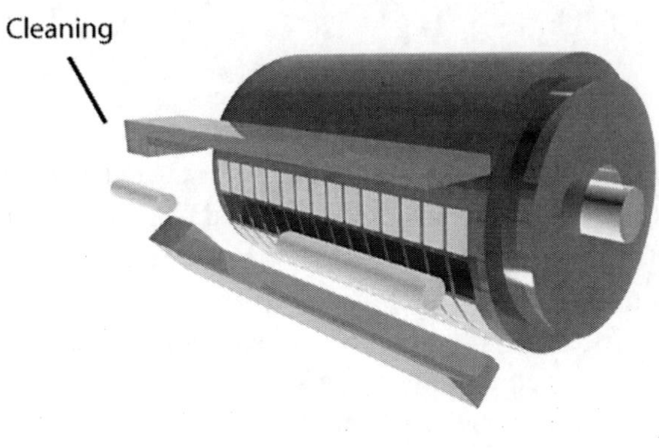

Figure 7.6 Cleaning

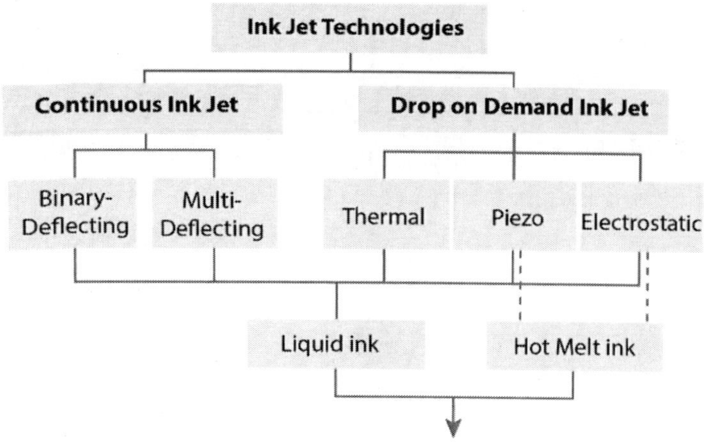

Figure 7.7 Inkjet technologies

onto the substrate during printing per the image signal. Instead, the unprinted droplets are collected and returned for reuse (Soleimani-Gorgani, 2015).

Binary deflection and multi-deflection are two ways of deflecting the drops in continuous inkjet. The beginning of the binary deflection CIJ printing process is connected with Carl Hellmuth Hertz from the Lund Institute of Technology in Sweden (in the 1960s). In opposition, R. Sweet from Stanford University was involved in multi-deflection CIJ (Heston, 2002; Keeling, 1981; Schneider and Hendricks, 1964; Sweet, 1971; Soleimani-Gorgani, 2015).

Figure 7.8 presents one example of the binary deflection process. The ink drop has one of two charge states. Uncharged states are for conveyance to the paper, and charged state is for deflection in an electrical field. That means the uncharged droplets are ejected directly to the substrate, while the charged droplets are recycled back into the printing system (Majnarić, 2015).

In the multi-deflection process, charged droplets are ejected directly to the substrate, but droplets receive different electrical charges and are diverted onto the substrate at different angles. Finally, the uncharged droplets return to the printing head (Figure 7.9).

Although their use is not widespread, Hertz and Microdot printing processes are also part of continuous inkjet.

Drop-on-demand inkjet technologies could be divided into three main groups: piezo inkjet printing process, thermal inkjet printing process, and electrostatic inkjet printing process.

The difference between these techniques is the way they generate a pressure pulse because in this process, as the name suggests, droplets are ejected from the nozzles by a pressure pulse created when necessary (Kyser and Sears, 1976; Stemme, 1973; Zoltan, 1972). For example, in the piezo inkjet printing process, the drop is generated by changing the chamber volume in a nozzle channel (Figure 7.10), while in the thermal inkjet process, the drop is generated by heat transfer. The droplets' ejection through the nozzles is induced by electrostatic forces characteristics of the electrostatic inkjet printing process.

The sequences in Figure 7.10 show in a simplified form how an ink drop can be ejected due to the ejection of an ink drop by mechanical displacement in the ink channel (Novaković and Kašiković, 2013).

The piezo inkjet printing process can be divided into a bend (Figure 7.11), push (Figure 7.12), shear (Figure 7.13), and squeeze mode (Figure 7.14) (Le, 1998).

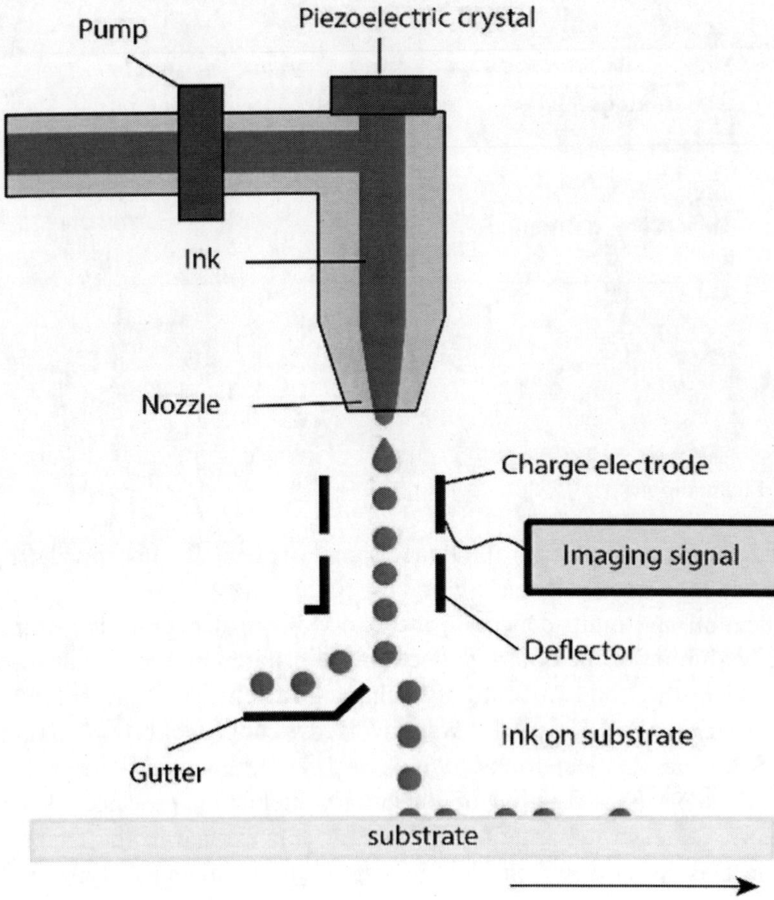

Figure 7.8 Binary deflection inkjet

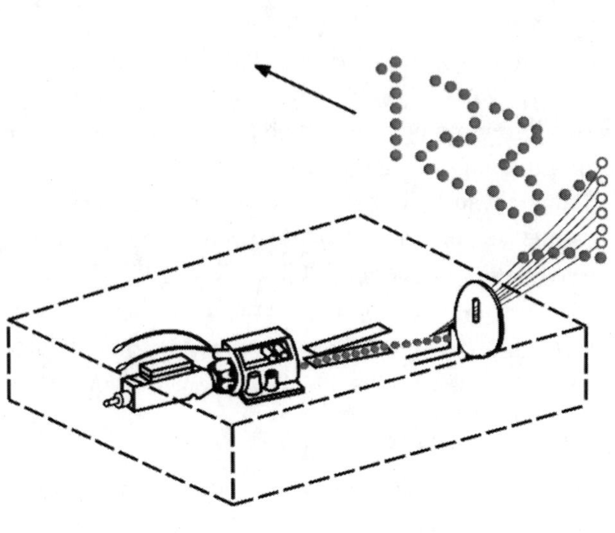

Figure 7.9 Multi-deflection inkjet

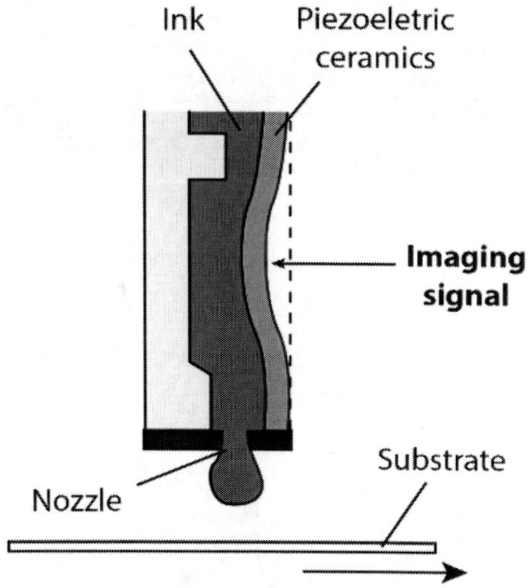

Figure 7.10 Piezo inkjet printing process

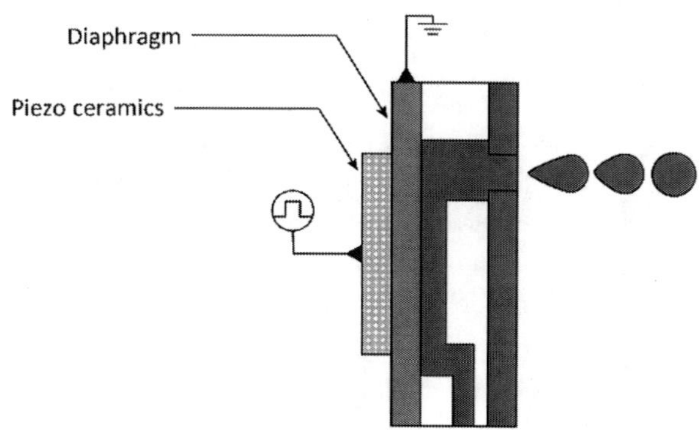

Figure 7.11 Piezo inkjet printing process – bend mode

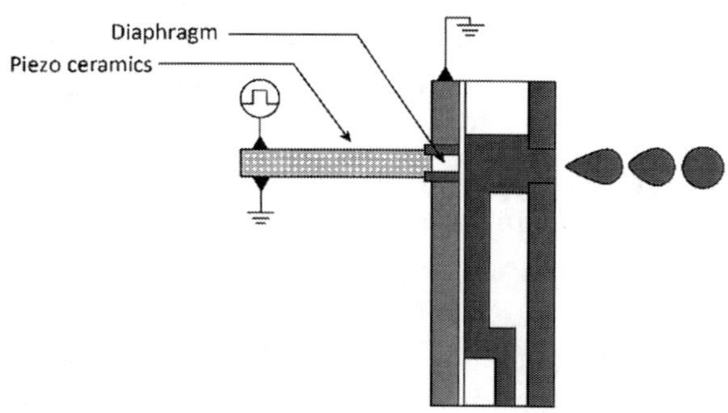

Figure 7.12 Piezo inkjet printing process – push mode

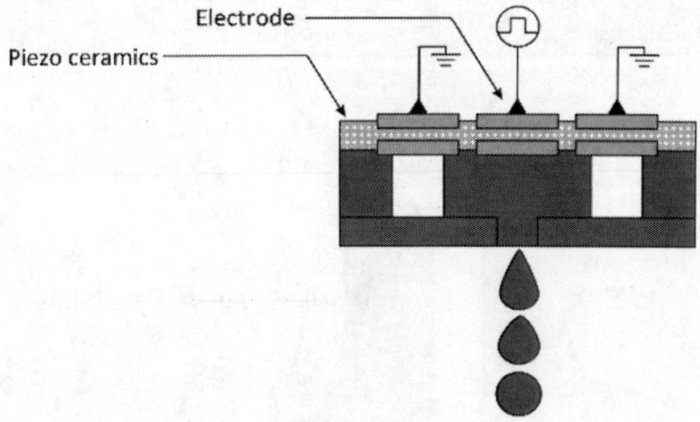

Figure 7.13 Piezo inkjet printing process – shear mode

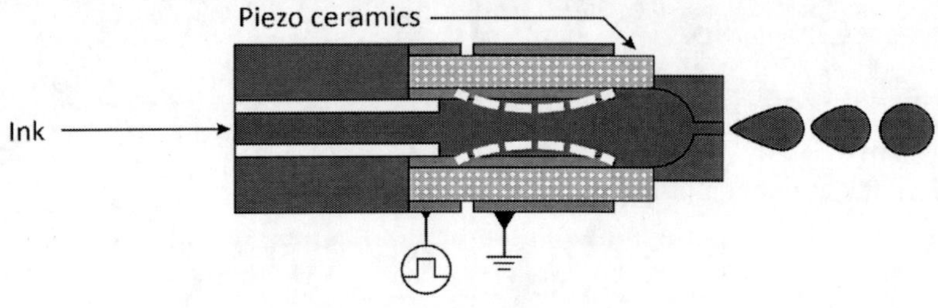

Figure 7.14 Piezo inkjet printing process – squeeze mode

Bend mode uses Tektronix, Epson, and Biofluidix printheads. In this process, the electric field and poling directions are parallel. The piezoelectric material is placed on the membrane, and the membrane moves like an oil can (Freire, 2006).

The characteristic of a push mode is that the electric field and polarization vectors are parallel. As a result, the membrane is placed in the expanding direction of the piezoelectric material. Hitachi, Dataproducts, Epson, and Biofluidix are using push mode in printhead production.

Shear mode is presented in Figure 7.13. Again, the electric field is perpendicular to the poling direction of the piezoelectric material. This mode is used by Tektronix, Sharp, and Xaar printheads.

In the squeeze mode, the drop ejector is a hollow tube of piezoelectric material (Figure 7.14). Upon applying an electric field, the inside volume of the tube (firing chamber) decreases its radius and ejects the ink in the direction of its axis (Freire, 2006). Squeeze mode is part of Microdrop and Microfab printheads production.

The print heads must be heat resistant for all these modes. Also, the spacing of droplets must be controlled by angling the line of nozzles along the print head, and last but not least, the droplets' speed, size, and shape can be controlled by adjusting the waveform of a voltage applied to the nozzles.

In the thermal inkjet printing process, an ink drop is generated by heating and vaporizing the inkjet system (Figure 7.15). In this process, a resistive element can form ink droplets within a few microseconds (the temperature is rapidly raised to 300–400°C). Bubble forces the ink out of the nozzles. It is created by high temperature, resulting in ink evaporation (Le, 1998).

Print heads in this process can create small-size droplets, and it is interesting that they are not expensive (Soleimani-Gorgani, 2015).

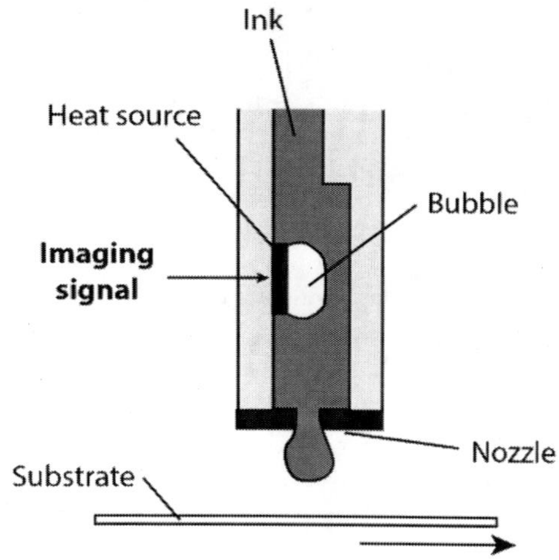

Figure 7.15 The thermal inkjet printing process

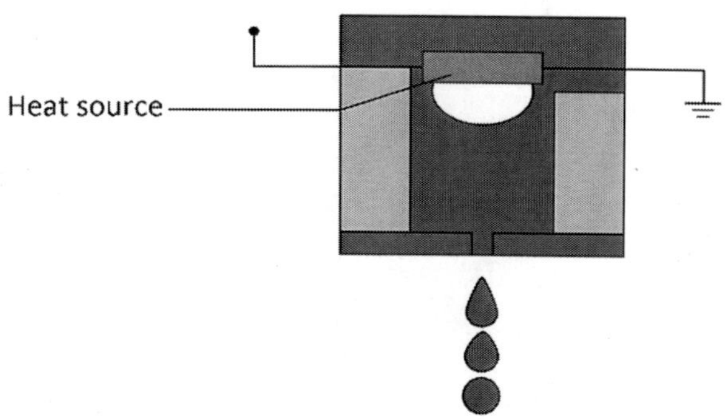

Figure 7.16 Thermal inkjet printing process – 'Roof-shooter'

'Roof-shooter' and 'side-shooter' types (depending on the heater position) are the most common configurations of drop ejectors. Hewlett Packard, Lexmark, and Olivetti use the 'roof-shooter' type of drop ejectors (Figure 7.16). The orifice is located on top of the heater.

The side shooter design is implemented in Canon and Xerox printheads. For example, in Figure 7.17, it is shown that the heater is located on a side close to the orifice (Le, 1998).

These are not the only methods for drop ejections. They can be found less often in 'back-shooter,' multi-heater, double heater, suspended heater, and movable member drop generator designs.

The electrostatic inkjet printing process is not popular as the previous two processes (Figure 7.18). Only a few companies, like TTP and Seiko Epson, use the electrostatic inkjet model for commercial applications (Soleimani-Gorgani, 2015). This process generates an electrical field between the nozzles and the substrate. The ink drops are produced by sending image-dependent control impulses to the nozzles. These impulses cause an ink drop to be released and routed through the electrical field onto the substrate (Kipphan, 2001).

Introduction to CAD Plotters

At the beginning, plotters differed from printers in image output technology. In other words, plotters drew and printers printed. The first plotters were devices for creating drawings, graphs,

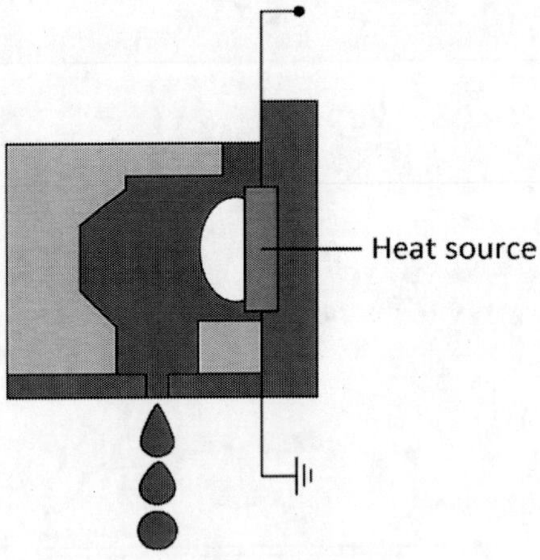

Figure 7.17 Thermal inkjet printing process – 'Side-shooter'

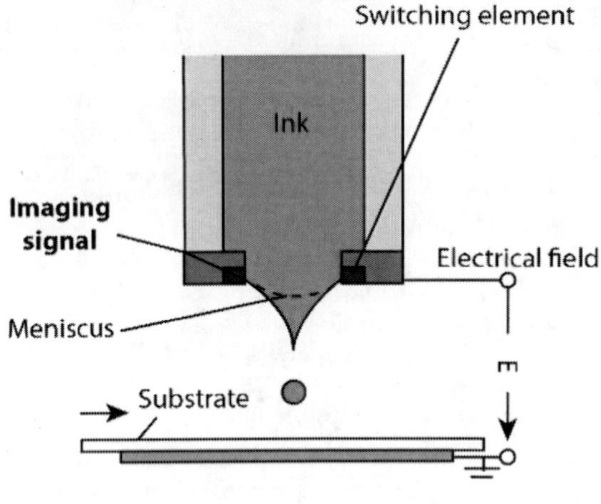

Figure 7.18 The electrostatic inkjet printing process

and diagrams on large-format sheets of paper (A1, A0). They were used mainly in engineering circles and were designed to replace the monotonous, slow, and manual labor of a draftsman. Nowadays, plotters also print; however, they are generally much more productive than conventional printers, have advanced features, print on large formats, and usually have dedicated memory. Apart from that, some models of plotters have a cutting function. This function is not available for conventional printers and, in addition to being used for large formats, is a property that differentiates plotters from printers.

Plotters allow users to perform printing on large formats and create drawings, 3D postcards, advertising signs, and other products. In addition to large advertising posters, they are indispensable for creating topographic maps, drawings, and projects. Regarding substrate, plotters can be both sheet-fed and roll-fed. Formats of the sheets are usually larger (i.e., A0, A1, or A2 format), and the width of the roll of the substrate can vary from 0.6 to 1.4 m.

Main Constructive Differences between the Conventional Printer and Plotter

Plotters and conventional printers have a lot in common, especially in the area of printing heads and print head positioning, meaning that the main differentiating factor between these two groups is the construction of the substrate-feeding system, especially for roll-fed plotters.

Conventional printers print mostly on paper. However, plotters can print on a large variety of substrates, i.e., cardboard, fabric, film, and other synthetic materials. Apart from that, some models can be used for printing directly on clothes (i.e., T-shirts) or even on some round objects, such as cups.

The substrate-feeding system has an encoder sensor in the form of a disk usually located on one side of the plotter. The disk has a large number of opaque strokes and transparent areas between them. Apart from the disk, there is an optical sensor comprised of emitting (LED) and light-receiving (photodiode) elements. The emitting element focuses the light through the encoder, which is later received by the photodiode. The light is blocked by the opaque strokes of the disk encoder. The larger the number of opaque strokes, which have crossed the ray of light means that more substrate was fed.

More modern plotters are equipped with a substrate-feeding mechanism with a stepper motor. Its rotor does not rotate continuously but takes on a series of discrete positions. When a control pulse is applied, the rotor rotates for a strictly defined angle, which can be quite small. This means that the substrate is fed step by step.

Types of Plotters by Functionality

Initially, plotters were intended only for printing purposes, but over time, they have been improved and modified to serve a few more purposes. All of that led to them being able to produce an even larger variety of products and take over some of the functions which previously required dedicated machines or devices (i.e., cutting). There are now three types of plotters that differ in functionality: cutting, printing, and combined plotters. Combined plotters are more commonly called hybrids.

Cutting Plotters

The working element of a cutting plotter is a knife with an optical sensor that ensures cutting accuracy. These plotters can cut out a complex pattern according to a path set by the program. Several parameters determine the cutting plotter quality:

- format
- slicing width
- cutting speed
- cutting accuracy (higher accuracy blade is needed for thinner substrates)
- knife pressure (higher pressure is needed to cut dense materials)
- type of knife

The knife pressure and type are the parameters that mostly determine which substrates are suitable for cutting on the cutting plotter. Different combinations of these parameters allow cutting plotters to be used on a large variety of substrates such as the following:

- self-adhesive and magnetic vinyl
- thermal films
- photo paper
- leather
- textiles
- plastic

Printing Plotters

Printing plotters can only transfer an image to the media. Instead of a knife, the working element of the printing plotter is a drawing block, and the ink transfer method depends on the used printing technology. Each drawing block is capable of working with a certain type of media.

Printing plotter quality is determined through a large number of properties:

- Print speed – Mostly affects performance and is not that important if the plotter is not meant to be used for printing large quantities of products.
- Resolution – Specifies the clarity of the printed image at a certain size.
- Format – Defines the maximum size of the substrate and image.
- Quality of color reproduction – Good color reproduction is important for printing advertising, but not so much for printing graphics and blueprints where color saturation and brightness usually do not matter.

Combined Plotters

Combined plotters, also known as hybrid plotters, have both cutting and printing plotter functionalities. They often use UV inks, and apart from process colors, they can print with White Ink and UV Clear Coat. Nowadays, hybrid plotters are devices that can be used to print, varnish, simulate embossing, crease, perforate, and contour cut a large variety of substrates. They are ideal for proofs, packaging prototypes, labels, and short-run production of personalized products. Using UV Clear Coat, they can achieve a glossy finish or even custom texturing or simulation of embossing.

These plotters are very capable and can print on foils, films, clear and metallic stock for labels, decals, displays and posters, natural and synthetic leather, window coverings, fabrics, and many more materials.

Hybrid plotters are usually capable of printing both on sheet substrates and substrates from rolls.

Types of Plotters according to the Substrate Form

The printing substrates are usually paper, cardboard, film, synthetic fabric, and other materials. They come in the form of sheets or rolls. Plotters are designed to print on sheets, rolls, and in newer models both sheets and rolls.

Sheet-Fed Plotters

Sheet-fed plotters are equipped with a table for supporting the substrate. They can print only on substrates in the form of sheets, most commonly A2 or A3. The substrate is permanently fixed on the table by magnetic, pneumatic, electrical, or mechanical fixation. To support the substrate, the table has to be large, meaning that they take up a lot of space and are therefore not very popular. They are most commonly used for the printing of large advertising signs.

Roll-Fed Plotters

The main advantage of a roll-fed plotter is compactness. The media is wound into a roll and fed along the drum to the drawing block, where in the process of moving, the drawing block applies the image. The device requires little space. Roll models save consumables and have better print quality and versatility than sheet-fed plotters.

Types of Plotters according to the Method of Creating an Image

Pen Plotters

The drawing block of pen plotters is a device equipped with elements in the form of pens. When printing, the block with the pen moves in two directions (X and Y axis). These plotters are classified as vector-type plotters, and they come in two varieties depending on the type of pen used for drawing:

- pens with liquid dye
- slate pens (pencils)

Although relatively slow and noisy, these plotters can achieve high print quality, with high levels of contrast and accurate color reproduction.

Inkjet Plotters

Regarding printing technology, these plotters resemble ordinary household inkjet printers. The printed image is generated using liquid ink, which is ejected through the nozzles of the printing head in the form of tiny droplets. The printing heads are equipped with a large number of nozzles that use different technologies and rely on different phenomena to form the droplet. Some of them heat up during printing, forming an air-filled bubble, which, under pressure, squeezes the ink from the nozzle. However, as the durability of inkjet printers mostly depends on the technology used for ejecting the ink from the printhead, modern CAD plotters have to rely on piezo technology for ink ejection. This means that they are not prone to heat-induced degradation, which is present in printheads based on thermal technology. Apart from increased durability, piezo printheads have the advantage of being able to use both dye and pigment-based inks, as well as inks that are susceptible to changes in properties when heated.

Piezo printheads are continuously developed. For example, Epson Micro Piezo printheads achieve superb placement accuracy with fast ejection of up to 50,000 droplets per second. Each nozzle can selectively eject ink droplets in a range of sizes. By precisely controlling the voltage waveform that drives the piezoelectric elements, it is now possible to eject ink droplets as small as 1.5 picoliters and deposit them exactly where needed (Epson, 2022).

Increasing piezoelectric element displacement is essential to selectively ejecting ink droplets of varying sizes. The thinner the piezoelectric elements, the greater their displacement under the same applied voltage. However, thinner piezoelectric elements are more fragile and thus hard to fabricate and handle. To solve the problem, Epson adopted a multi-layer piezo, consisting of a stack of thin piezoelectric elements as an actuator, and afterward developed a multi-layer actuator head to eject ink through vertical vibration of the piezoelectric elements (Epson, 2022).

Epson Micro Piezo printheads also use Active Meniscus Control, meaning that the ink that's left in the nozzle, as a droplet leaves the nozzle, has its oscillations dampened by intentional counteracting movements of the piezo device. The ink at the nozzle surface remains essentially flat across the opening, and largely devoid of motions that would otherwise cause the next droplet to emerge off course or lose its nearly perfect spherical shape (Joseph Holmes, 2007).

Inkjet plotters are commonly used for color printing and the production of outdoor advertising. They are not limited to printing only on paper. Substrates such as banner fabric, canvas, or self-adhesive film are also easily printed. They have relatively high productivity, are simple to operate, achieve good image clarity, and the consumables are generally available.

Toner-Based Plotters

Toner-based (laser and LED) plotters are based on electrographic technology, meaning that they use toner as a dye. Furthermore, in laser-based models, the laser beam can be used to create a latent image on a drum and to cut the prints. Both laser and LED technology rely on light-sensitive semiconductor layers (usually with selenium) and electrostatic field effects. A selenium-coated drum serves as an intermediate carrier of the image and is capable of being charged to high potential. Afterward, a laser or LED light beam removes the charge from areas where toner particles should be attracted, thus creating a latent image. After being transferred to the substrate, toner and substrate pass through the fixing unit, where the toner gets baked into a permanent image.

The main difference between laser and LED toner-based plotters is in the light source. While laser plotters use systems of lenses to move and focus a beam over certain areas of the drum, LED plotters use small LEDs placed linearly across the entire drum width. The number of LEDs is high, usually more than 10,000, and they can be turned on or off individually, thus creating high-resolution latent images on the drum extremely fast.

LED plotters are considered more advanced than laser plotters. However, both technologies have proven to be useful for printing drawings, diagrams, and other graphics where accuracy, high resolution, and image clarity are of utmost importance. Contrary to other technologies used for printing plotters, especially inkjet, laser and LED toner-based plotter heads are not prone to wearing out due to the ink drying and clogging the nozzles of the printing head. Toner-based plotters are very fast, and that is one of the main characteristics which made them popular both for advertising and printing in the domain of engineering. The main downside of this technology is the high cost of the equipment, especially nowadays when inkjet plotters have been further developed and modernized.

Solid Ink Plotters

Solid ink plotters have a fixed print head and a rotating drum. The drum has to be cleaned from ink residues before each cycle. This is done automatically by applying a special lubricant to the drum surface. Ink solidifies on the drum due to a large temperature difference between the drum and the print head. Afterward, the roller is used to transfer the ink to the substrate following the same temperature difference principle. Solid ink plotters were mostly aimed at the designers and are now the technology of the past, despite having good characteristics such as:

- sustainable and repeatable print quality;
- relatively low cost; and
- being suitable for printing glossy photographic images that do not lose their appearance and clarity.

Electrostatic Plotters

An electrostatic plotter is somewhat similar to a toner-based plotter. Its principle of operation is also based on the formation of latent images. Unfortunately, the disadvantage of this technology is that printing requires special electrostatic paper, which needs to be thoroughly coated with a dielectric and impregnated with a hydrophilic salt. Nevertheless, electrostatic plotters can quickly print images and drawings of any level of complexity. Products printed using electrostatic plotters have proven to be quite stable and resistant to external factors. Prints are resistant to UV rays and do not lose brightness after being stored for longer periods. Apart from being resistant to UV rays, prints obtained using electrostatic printers also have a high degree of protection from other external climatic conditions.

However, the maintenance of electrostatic plotters is complex, and the consumables price is not competitive. Electrostatic plotters work properly only if placed in a room with recommended humidity and air temperature. Electrostatic papers also require constant storage conditions, especially regarding the humidity level, as even the slightest oscillations in humidity can cause permanent damage to the paper's dielectric coating.

Direct Output Plotters

Similar to electrostatic plotters, direct output plotters also require special thermal paper. Thermal papers are also specially coated and impregnated. They are fed from a roll and pass through a special head – a comb. A comb is equipped with a large number of heating elements, which, depending on the temperature, causes black color or certain shades of gray to occur on heated areas of thermal paper. Contrary to the electrostatic plotters, substrates and other consumables for direct output plotters are much more widespread and available. Substrates are available both in the form of papers and films.

This technique produces high-definition black-and-white images, making direct output plotters popular among engineering organizations for printing projects. This property makes them similar to some modern inkjet plotter technologies that use up to three shades of gray to obtain high-quality technical prints. Finished products obtained using direct output plotters also have a long shelf life and minimal sensitivity to external influences.

Thermal Transfer Plotters

The printing head (comb) of thermal transfer plotters is the same as the head of direct image output plotters. They also require special thermal paper as a substrate and a special film called a color donor placed between the comb and the paper. The thickness of the color donor usually does not exceed 4–10 microns. During printing, the comb passes over the donor, causing the donor to warm up and leave a pattern of the desired color on the thermal paper.

Most often, thermal plotters are used where it is necessary to create an image with high color reproduction quality, for example, in cartography. Printed products are moisture resistant and do not fade when exposed to UV rays. The main disadvantages of thermal transfer plotters are the high cost of consumables and the need for frequent replacement.

Sublimation Plotters

Sublimation plotters work on the principle of obtaining saturated shades under the influence of high temperatures. They are most commonly used for large-format printing of artistic and photographic images on textile and polymeric materials. High temperatures cause pores of the substrate to open up, thus allowing more ink to penetrate into them and be better fixed.

Some products printed using sublimation plotters are tapestries, flags, curtains, upholstery fabrics, T-shirts, sportswear and uniforms, scenery for theater and television, and even glass. Despite being very versatile, there are still some limitations regarding the substrates. For example, synthetic materials are suitable for printing, but natural ones are not.

Sublimation printing is outstanding for fine lines and details. Set-up costs are low, making it good for shorter runs. It is possible to easily achieve realistic color reproduction and permanent quality, as the prints are extremely resistant to UV rays.

Types of Inks Used in Plotters

Water-Based Inks

Water-based or aqueous inks come in the form of aqueous dye inks and aqueous pigment inks. In aqueous dye inks, a colorant is a dye that is water soluble. During printing, the ink is absorbed by the printing substrate, most commonly paper. Plotters using this kind of ink can produce extremely fine lines and easily print detailed technical drawings. In aqueous pigment inks, colored pigment particles are suspended in water. The water evaporates during drying, leaving the pigment particles adhered to the substrate surface. It is important to note that aqueous pigment inks cannot be used with the plotters which use aqueous dye inks and vice versa.

Solvent-Based Inks

Instead of water, solvent inks contain solvents, which allow for printing on PVC surfaces. They contain light hydrocarbons, which are isolated from oil and coal. The technology has proven to be good as it produces waterproof prints and prints resistant to the effects of UV rays and low temperatures.

Solvent plotters are mainly used for color printing of outdoor advertising, posters, photographs, banners, and other types of printed products. They are also used for printing floor coverings and wall panels made of plastic, polystyrene, PVC, and other materials.

Eco-solvent Inks

Environmental problems have led to the fact that many manufacturing companies have begun to produce plotter models using eco-solvent inks. They also contain solvents but are less aggressive and toxic.

They are practically odorless and less harmful to the environment. Another advantage is that the pigment particles are small, thus providing high resolution of prints, even up to 1,440 dpi.

UV Inks

UV inks are cured using the UV beams, meaning that the plotters which print using UV inks have a UV light source attached to the printhead. When exposed to UV light, the ink forms a strong film on the surface of the substrate, making high-quality prints that are resistant to external influences.

UV plotters are equipment for large-format printing on rigid and flexible substrates such as plastic, metal, glass, ceramic tiles, cardboard, wood, fabric, leather, and other materials. They are widely used for printing advertising and souvenir products, decorative interior elements, and printing proofs.

The main disadvantages of UV inks and UV plotters are increased consumption of electricity, odor, nozzle clogging when not used, and the need to contain UV inks under certain climatic conditions.

Latex Inks

Special latex inks use completely harmless dyes that do not emit an unpleasant odor. Latex ink plotters are used for large-format printing on various substrates such as paper, self-adhesive and vinyl films, fabric, canvas, leather, and other materials. Some of the most common products printed using latex inks are wallpapers, bed linen, curtains, and other interior decoration items.

Table 7.1 shows an overview of different CAD plotter types' main advantages and disadvantages, following the classification based on ink type, while Table 7.2 shows the same overview based on image-creating methods.

Conclusion

Toner-based and inkjet plotters are the ones that are the most widespread. In the past, the break-even point between toner-based and inkjet plotters was around 200 m^2 per month of output. Nowadays, ink prices have dropped, and inkjet systems became much faster, especially with the introduction of single-pass inkjet systems. This means that the operational cost of inkjet printing for line drawings is now less than the average cost of printing with toner. Apart from that, the average print runs are getting lower, further leading to the inkjet plotters being fast enough in most of the cases.

Table 7.1 Overview of main advantages and disadvantages of different CAD plotter types regarding used ink

Image-creating method/ Ink type	Advantages	Disadvantages
Water-based inks	Eco-friendly; high-quality prints	Can be used only on paper/cardboard substrate
Solvent-based inks	Resistant prints; suitable for printing on foils and films	Safety precautions needed for working with solvents
UV inks	High quality; can be used on various materials; resistant prints	Printing head maintenance; control and maintenance of ink

Table 7.2 Overview of main advantages and disadvantages of different CAD plotter types regarding image-creating method

Image-creating method/ Ink type	Advantages	Disadvantages
Pen plotters	High levels of contrast; accurate color reproduction	Outdated; slow and noisy
Inkjet plotters	Low cost for color prints	Requires coated paper to prevent fading
Toner-based plotters	Easy maintenance; high quality	Has to be installed on a perfectly flat floor; expensive equipment
Solid ink plotters	Fast; high quality	Ink storage requirements
Electrostatic plotters	Fast; prints resistant to most of the external factors	Complex maintenance; price of consumables not competitive
Direct output plotters	Long shelf life of products; prints resistant to external factors	Can print only in black and white; requires special paper
Thermal transfer plotters	High color reproduction quality; UV and moisture resistant prints	Requires special paper and color donor; high cost of consumables
Sublimation plotters	Realistic colors; good contrast	Not suitable for natural fabrics

Inkjet plotters are not designed to be long-lasting, in contrast to toner-based plotters, which can last more than ten years, especially in the case of LED toner-based plotters.

Some of the inkjet plotters can print up to 1,000 m^2 per month, but a large number of them are rated only for around 300 m^2 per month. On the contrary, most toner-based plotters can easily print more than 2,000 m^2 per month.

As for the printing speed, modern high-end inkjet plotters are three to five times slower in comparison to average or high-end toner-based plotters. However, with the introduction of single-pass technology, which employs a long, stationary print head that ejects ink onto a traveling substrate underneath, the speeds went up to four times in comparison to the standard inkjet plotters with the traveling print head.

References

Epson (2022) *Micro Piezo Inkjet Technology.* Available from: https://global.epson.com/technology/printer_inkjet/micro_piezo.html [Accessed 15 July 2022].

Freire M. (2006). *Inkjet Printing Technology (CIJ/DOD)* In Ujiie H. (ed.), *Digital Printing of Textiles.* Cambridge, Woodhead Publishing Limited, pp. 29–52.

Heston S.F. (2002). *CIJ and Experimental Setup Schematics.* Pittsburgh, Pennsylvania, USA: University of Pittsburgh.

Joseph Holmes (2007). *EPSON's Next Generation – The World Leaders in Print Quality Go Super Wide, While Introducing Lots of Improved Technology.* Available from: shorturl.at/chtuZ [Accessed: 15 July 2022.]

Kašiković N., Novaković D., Juric I. (2016). *Digitalna štampa – praktikum za vežbe.* Novi Sad, Serbia, Faculty of Technical Sciences.

Keeling M.R. (1981). Inkjet printing. *Physics in Technology* 12(5): 196.

Kipphan H. (2001). *Handbook of Print Media, Technologies and Production Methods.* Berlin, Springer.

Kyser E.L., Sears S.B. (1976). *Method and Apparatus for Recording with Writing Fluids and Drop Projection Means Therefore.* US patent 3946398.

Le H.P. (1998). Progress and trends in inkjet printing technology. *Journal of Imaging Science and Technology* 42: 49–62.

Majnarić I. (2015). *Osnove digitalnog tiska.* Zagreb, Croatia, Sveuciliste u Zagrebu.

Novaković D., Kašiković N. (2013). *Digitalna štampa.* Novi Sad, Serbia, Faculty of Technical Sciences.

Océ (2006). *Digital Printing - Technology and Printing Techniques of Océ Digital Printing Presses.* Poing, Océ Printing Systems GmbH.

Schneider J.M., Hendricks C.D. (1964). Source of uniform-sized liquid droplets. *AIP Review of Scientific Instruments* 35: 1349.

Soleimani-Gorgani A. (2015). *Inkjet Printing.* In Izdebska J., Thomas S. (eds.), *Printing on Polymers: Fundamentals and Applications.* Oxford, Elsevier, pp. 231–241.

Stemme N. (1973). *Arrangement of Writing Mechanisms for Writing on Paper with a Colored Liquid.* US patent 3747120.

Sweet R.G. (1971). *Fluid Droplet Recorder.* US patent 3596275.

Zoltan S.I. (1972). *Pulsed Droplet Ejecting System.* US patent 3683212.

Chapter 8

New Concept Design and Vibromechanical Analysis of a Traditional Greek String Musical Instrument

Ioannis Pimenidis,[1] Christos Koidis,[1] Panagiotis Kyratsis,[2] Dimitrios Tzetzis[1,*]

[1] Digital Manufacturing and Materials Characterization Laboratory, School of Science and Technology, International Hellenic University, Thessaloniki, Greece

[2] Department of Product and Systems Design Engineering, University of Western Macedonia, Kila, Kozani, Greece

Abstract

Traditional music's contribution to people's culture is extremely important. Its distinct sound color springs out from the people but simultaneously emits the people's attitude to life. The austere but substantial form of musical instruments is usually conceived and designed intuitively. As this process is not well-defined, the range of the performance of the instruments is very wide, even among the instruments, made by the same luthier. Thus, the need for standardization of traditional musical instruments manufacturing processes is quite significant. In that context, the lyre of Pontus was redesigned via contemporary production methods and its properties were investigated. With respect to tradition, some improvements are proposed and modeled with the use of Computer-Aided Design and Computer-Aided Engineering software. The latter is also employed for the investigation of the rigidness and the vibrational behavior of the final concept. Prior to that, a material selection methodology, based on specific criteria that concern the attributes of the material for the particular use, proposed the materials which are more suitable for that task.

Keywords: Pontus, lyre, 3D model, carbon fiber-reinforced plastic, material selection, mechanical analysis, standardization

Introduction

Traditional music is a very important part of a nation's culture. Its content reflects people's customs and habits but more than that, along with its rhythm, its color, and its mood, it reveals people's worldviews and attitude to life. The technology used to produce the traditional musical

* Corresponding Author's Email: d.tzetzis@ihu.edu.gr

instruments was based on the experience of the makers while the materials selected were the most simple and easy to find. The combination of makers' experience in manufacturing with the use of traditional materials with optimized properties enabled the creation of simple in implementation but great in design musical instruments. Experienced users have the skills to detect the qualitative differences of several instruments empirically. The typical method of testing is to hear the instrument but mostly to interact with it. Novice players usually do not have that ability. In both cases, the examination of the instrument via physical contact is obligatory as there is not a robust standardization method involved in the manufacturing process of the instrument that ensures its quality. Even the same luthier may produce musical instruments that perform differently.

The name "lyre" refers directly to the ancient Greek lyre. Although the differences between the pontic and the ancient Greek lyre are obvious, the emotional relation is quite strong. It is difficult to trace with certainty the exact evolution line of the pontic lyre. There is not any hard evidence that can record who, when, what, and how, regarding the instrument. Only, intuitively indirect conclusions can be made. Regarding the lyre's body, it can be said that the pontic lyre is a crossbreed of the ancient Greek lyre, Pythagoras's monochord, and the ancient pandouris [1, 2]. More analytically, the vertical orientation of the lyre, while performing, is like the orientation of the ancient Greek lyre. Furthermore, like the ancient lyre, the pontic one does not have any frets. The sound box of the lyre has a quite archetypical form, which is close to Pythagoras's monochord sound box. The instrument has three strings and it is all made of wood, like the pandouris or trichordon (the compound word pandouris, "pan-" means whole and "-dori" means wood, expresses that quality).

In that sense, redesigning traditional musical instruments such as the pontic lyre is not a simple task. The shape, the sound, and the way of performing the lyre invite the listeners to be a part of an exceptional experience. It is not an exaggeration to say that listening to this kind of music and participating in a pontic happening is an experiential process that may lead, in some cases, to a state of sacred methexis. The pontic lyre has a shape that is unique in the sector of traditional instruments in Greece. While all the other lyres in Greece (Cretan, Thracian, Macedonian, Constantinopolitan, and insular) are pear shaped, the lyre of Pontus is bottle shaped.

Pontic Lyre Design

The body of the lyre is wider at the bottom, and it is getting narrower as the eye moves upwards as it is shown in Figure 8.1 [3]. On the top of the lyre, there is the head. The head has three holes where three pegs, which are used for the strings' tuning, are inserted, and wedged [3]. A fretless fingerboard is placed on the top of the instrument's neck and just under the head. A few centimeters under the fingerboard is placed the bridge, a small but very crucial component for the lyre's sound performance. The bridge transfers the vibrations that the strings produce to the soundboard (lid) of the instrument. The holes on the ribs of the instrument and on the soundboard, along with the curved slots (nostrils), allow the sound to come out of the instrument. On the low part of the instrument, there is the string holder. This part is anchored underneath the instrument with a small wooden hook and a piece of wire. Underneath the right column of the bridge is a small sound post (stoular). The sound post exists in all bowed instruments and its absence decreases the performance of the instrument to 20% [3]. It transmits the strings'

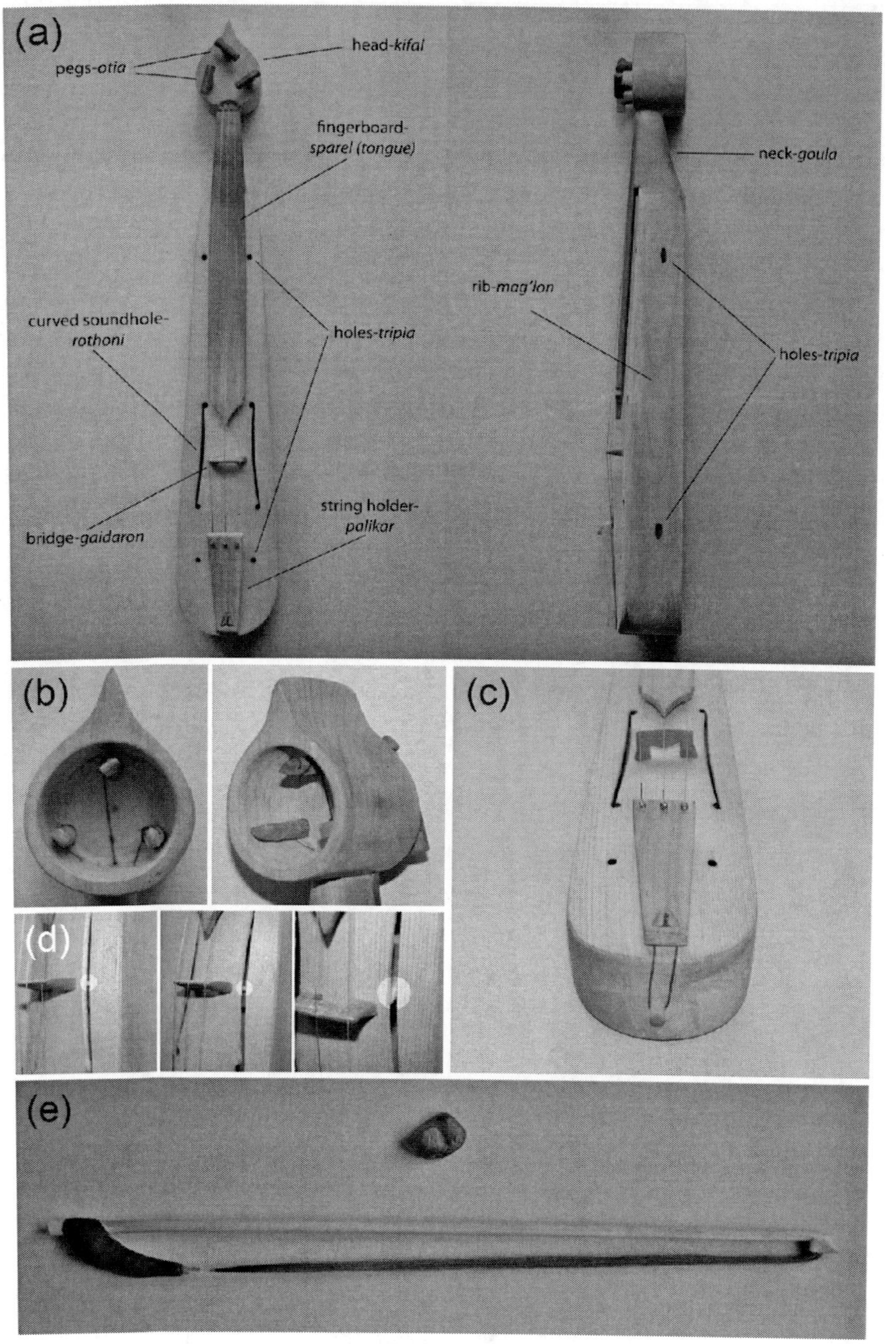

Figure 8.1 (a) The anatomy of the lyre (front and side view), (b) the tuning pegs of the lyre (back view), (c) the anchoring of the string holder, (d) the sound post (stoular) of the lyre, and (e) the bow and its resin.

vibrations to the back of the instrument and tones up the difference between the low (bass) and the high (treble) sounds of the lyre [4]. The importance of this component is emphasized by the fact that it is also called "the soul" (psyn) of the lyre. In Karpathos island, Greece, it is called "the wood of the Christ" ("Christoksilo") [5]. The back side of the instrument is flat. The back side of some old, handmade instruments, although flat outside, was curved inside to allow, this way, the sound to exit in a more effective way through the lyre's openings [3]. Finally, the bow of the lyre is a simple piece of wood, which has a bunch of male horse's tail hair attached to it (via a loose manner). A piece of natural resin is applied to the bow's hair to increase the friction with the strings [3]. Three variations of the pontic lyre exist: (1) kapan, (2) zil, and (3) zilokapani [3]. They all share the same shape, while differing in dimensions. This results in different sound colors. The first type is the most common. Its low tone (bass) produces a full, epic sound effect. The zil type is smaller and it is tuned in a high pitch, which results in a more vivid music color, as it is shown in Figure 8.2 [3, 6]. The kapan type is said to be more appropriate for singing, while the zil type for dancing [4]. The zilokapani type stands somewhere in the middle. A relative instrument to the pontic lyre is the "kemanes" [7] and "kemencello" [8].

The performance of playing the pontic lyre is affected by three main factors (the setting is for right-handed players, while for left-handed players the usage of the hands is inverted): (1) the pressing of the strings with the left hand, (2) the movement of the bow with the right hand, and (3) the swinging of the lyre also with the left hand. In all other types of lyres, the fingertips touch the fingerboard, and the strings are touched, to the side, with the nails. However, the pontic lyre is the only one whose strings are pressed with the fingertips. This technique allows the player to press simultaneously two strings and thus the characteristic polyphonic sound of the pontic lyre is produced. To achieve that, the space between the chords should be small enough to allow the simultaneous pressing of the two chords, but also large enough to provide the needed space for monochord playing. The double-chord playing characteristic allows the player to use

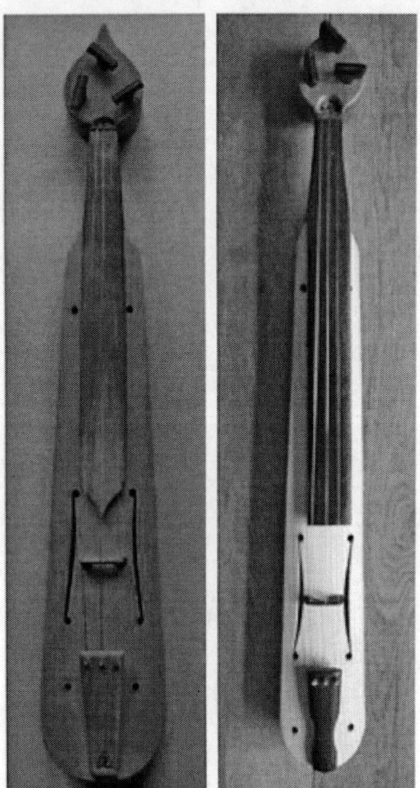

Figure 8.2 The "kapan" (left) and the "zil" (right) lyre types.

several playing motifs, even in the same song. The combinations may include monochord playing, double-chord playing, melody playing in the high (treble) string while keeping a steady tone (isokratis) in the low (bass) string, or melody playing in the low string while keeping a steady tone in the high string.

The bowing technique is at least equally important to the string's pressing techniques. The hand that holds the bow does not have a pathetic role. On the contrary, the bowing technique gives the tone to the melody and colorizes the music, by affecting the pace of the tense of the playing. During performing the lyre, the upper arm stays close (it is almost stuck) to the player's torso, while the forearm moves the bow back and forth. However, it is the movement of the wrist that gives the appropriate musical detail to the lyre's sound outcome. Finally, the lyre is not held steady like a violin. The swinging of the instrument with the left hand is synchronized with the movement of the right hand (bowing). Thus, the right hand does not need to make expanded movements, and the player can exploit the movement of the wrist. The whole technique is supported and enabled by the small curvature of the bridge.

Lyre Traditional Manufacturing

The small parts of the lyre play a very important role in its performance and their manufacturing is quite simple. The traditional manufacturing process is concentrated on the development of the two main elements of the lyre, the sound box and the soundboard because their production requires special knowledge and skills. Regarding the sound box, there are two approaches that categorize the instrument into two classes, the monoblock (chiseled) and the assembled type, so the instrument is referred to as a "chiseled lyre" or as an "assembled lyre." Opinions differ on which type is the best. In brief, a chiseled lyre is made of a single piece of wood that has uniformly distributed its sound properties. In contradiction, an assembled lyre is made by those parts of the wood that maximize the sound's transfer and its amplification. The methodology, which results in these outcomes, depends on the luthier (every luthier has his own manufacturing "secrets") and the tools she/he uses. The first step in both cases is the selection of the material [9–12]. One of the most popular materials for the manufacturing of musical instruments is wood. The selected wood should have as much vibratility and density as possible, and it should be well dried (low humidity) and stored at a low temperature. These properties minimize the absorption of the sound [13]. Also, the wood should be flawless, uniform, and straight grained (radial section) and with narrow growth rings (maximum 2 mm). These properties contribute to the optimization of the resonance effect. Furthermore, the wood should be cleaved and not sawn to have a straight-grained piece of material [14]. The traditional way to manufacture a lyre includes the removal and/or structuring of the material using a set of chisels and/or using mechanical means [15, 16].

Based on the above, the process for manufacturing the pontic lyre is based mainly on the experience and intuition of the luthier. The outcome of traditional manufacturing can be evaluated on two levels, affecting each other. First, the sonic result in conjunction with the technical detailing and playing easiness, and second the aesthetical detailing. The level of achievement in technical detailing defines the proper functioning of the instrument. For instance, if the keys are not robustly made, the lyre will be out of tune while playing. Furthermore, if they cannot be easily wedged to the body, then the instrument will not be able to be tuned in the desired tone. Also, the dimensioning of the instrument is a matter that must be tackled. The positioning of the lyre's parts and the openings on the lyre along with the thickness of the sound box's sides is a correlated subject. The wrong dimensions will affect the sound quality.

Regarding the playing easiness, the luthier should take account of the hand's ergonomics to allow performing the instrument easily. The neck should be fully grabbed with the palm of the hand and the distance between the strings should allow the single or simultaneous pressing of them by the player's fingertips. Also, the weight distribution of the instrument affects its playing easiness. An unbalanced lyre (high-weight center) will not allow the proper swinging during playing and therefore the movements of the two hands will not be well synchronized.

Although the aesthetical detailing may not concern the functioning of the instrument, it plays a key emotional role. The luthier should decide what stylish line to follow to highlight the importance (both historical and qualitative) of the instrument, without overloading it with unnecessary cosmetic elements. Also, the wood selection affects the appearance of the lyre. Wooden instruments have a natural look, and, in a sense, they are alive, since their performance is related to the environmental conditions and their treatment by the musician.

Finally, even though sound is the ultimate criterion for the instrument's quality, a well-made instrument should also be a piece of art. Traditional manufacturing provides a lot of choices in both directions, technical and aesthetical. Conclusively, as it is commonly said, no lyre is the same as another, even if they are manufactured by the same luthier.

New Pontic Lyre Design

For the new concept, the dimensions proposed by [17] are employed. According to this approach, the physical dimensions of the lyre encapsulate ratios that are derived from the lyre's tuning and playing. The fractions 3/4, 2/3, 9/16, 5/8, 3/5, 8/15, and 1/2 are used as dimensional ratios. The expected outcome of this method is to make the sound box vibrate in frequencies that are related to the lyre's tuning and playing, and consequently, to amplify the sound according to the characteristic acoustics of the pontic lyre. The string's length (from the base of the head up to the bridge) should be two-thirds of the instrument's total length [17, 18]. Consequently, as an A (La) tone is produced by a violin's string at a $V = 32.4$ cm length (approx.), the total length (L) will be $32.4/(2/3) = 48.6$ cm. The length from the bridge to the bottom of the sound box (D) will be $48.6 - 32.4 = 16.2$ cm. Regarding the sound box, according to [17], D is divided by 3/4 (perfect fourth ratio). $U = 16.2/(3/4) = 21.6$ cm is the length from the bridge to the top of the sound box. The total length of the sound box will be $16.2 + 21.6 = 37.8$ cm. Also, $U/V = 21.6/32.4 = 2/3$ (perfect fifth). The neck's length will be $32.4 - 21.6 = 10.8$ cm. The maximum width (W_{max}) of the sound box will be one-fourth of its total length. Therefore, $W_{max} = 37.8/4 = 9.5$ cm (approx.). The minimum width (M_{min}) will be three-fifths of W_{max}, that is, $(9.5 \times 3)/5 = 5.7$ cm.

The maximum height (H_{max}) will be 5.7 cm. The minimum height (H_{min}) will be three-fourths (perfect fourth) of H_{max}, that is, $(3 \times 5.7)/4 = 4.3$ cm (approx.). The sound box has four holes that enable the sound to come out of it. The holes create air pressure points on the sound wave and the harmonic's frequency with the maximum width at this point is heard [17]. So, it is chosen to open the holes on the 1/6 (minor third) of the sound box on the soundboard ($37.8/6 = 6.3$ cm) and on the 1/5 (major third) of the sound box on its sides ($37.8/5 = 7.5$ cm, approx.).

The cross section of the new concept's sound box was designed in Solidworks™ Computer-Aided Design (CAD) software as the intersection of two parabolas and is shown in Figure 8.3. This design enables the sound to come out smoothly from the instrument [4] and forces the

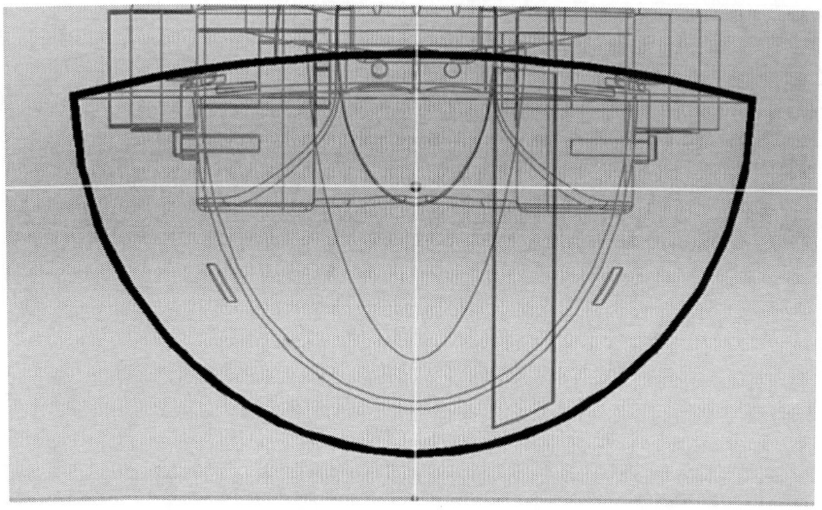

Figure 8.3 Cross section of the new concept (front view).

sides of the lyre to vibrate more vigorously. In a traditional lyre, the string, through the bridge, vibrates the soundboard. This vibration, through the sound post, is transmitted to the lyre's back. When the vibrations meet the sides of the instrument, they return to their source [17]. In the new concept, the sides of the lyre cannot be distinguished from the rest of the body, and consequently, the whole sound box is expected to vibrate in a more dynamic way.

The sound post of the lyre in a traditional instrument is flattened to be installed through the nostrils of the lyre. In the new concept, it has the same cross section as the bridge's leg to absorb the whole vibration as shown in Figure 8.4a. Also, the new form of the sound post is expected to vibrate mostly vertically (due to the lack of vibratility). Therefore, the major percentage of the vibrations is expected to be transmitted to the back of the instrument while the vibrations' losses will be minimized.

The sound box's thickness was chosen to be 1 mm to vibrate more efficiently. However, the bottom's thickness gradually is maximized to 3 mm to add weight and to help balance the instrument. Regarding the pegs, the classic pegs of an acoustic guitar were attached to an asymmetric petal-like pegbox (Figure 8.4b). The chosen pegs offer the needed tuning precision, in a simple operating manner and at a low cost. The pegbox allows the orientation of the pegs to be like the traditional instrument (vertical to the soundboard, in the forward direction). The pattern of the chosen string holder, as illustrated in Figure 8.4c, provides a robust geometry, minimizing the used quantity of material. An overview of the concept of the pontic lyre (with chords) is shown in Figure 8.4d.

Material Selection for the New Pontic Lyre

As it was mentioned before the material should be, primarily, vibratile. This property of the material will allow sound waves, which are produced by the strings, to vibrate the sound box-soundboard system. However, the instrument should recover to its initial dimensional state after the vibrations' depreciation. Consequently, it should be rigid, and, thus, the elasticity factor (Young's modulus) should be maximized. As the sound box will vibrate, it will press and

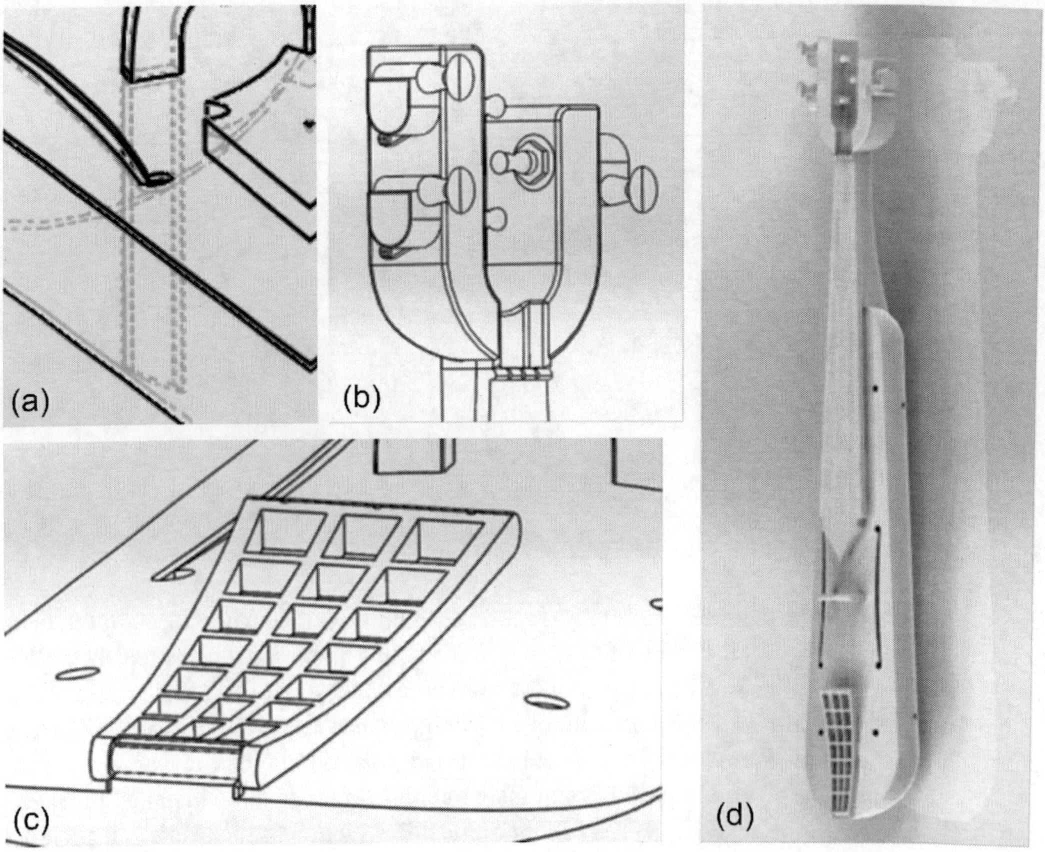

Figure 8.4 (a) The sound post of the new lyre, (b) pegs and pegbox, (c) the string holder, and (d) the new lyre with chords.

decompress the enclosed air. The produced sound wave should not be absorbed by the material. To minimize sound absorption, the material should be as dense as possible (non-porous). Therefore, the density should be maximized. The thermal expansion factor is another parameter that should be considered. Changes in temperature cause dimensional changes to the material. In the case of the lyre, this effect will result in the alteration of the sound color. Consequently, not only distortion is not acceptable but also thermal expansion should be minimized. This means that the material should have a small thermal expansion coefficient. Furthermore, as the instrument is under pressure by the strings, the selected material should tolerate as much compressive strength as possible, so as not to be deformed. This phenomenon will become more intensive if temperature change occurs. As it has been reported, the materials under thermal change are not stress free [19].

Materials such as ceramics, glasses, metals, alloys, foams, natural materials, and elastomers were omitted from the selection process. The use of these materials does not allow the cost-efficient production of an instrument such as a lyre, because either they require expensive production methods (while the sonic result of the product will be questionable) or they are not dimensionally stable. Carbon fiber-reinforced plastic (CFRP) is by far the more rigid material, while in terms of density all the materials are quite close to each other. The CFRP remains the best choice, since it undertakes maximum compressive load while it has a minimum thermal expansion coefficient. To compare the properties of the CFRP with the properties of the traditional materials that are used to produce musical instruments such as the lyre, the selection of the material was based on an analysis using the CES software. CFRP, plastics, and woods were compared, and the results confirmed that CFRP is the best choice (Figures 8.5 and 8.6). Some woods are more rigid and more thermally unaffected than plastics, while they can bear similar compressive forces. This fact gives them a comparative advantage against plastics. However, they are less dense, which may affect the quality of the sound that they produce, while some of them such as bamboo or plywood cannot be easily reformed into a sound box or soundboard.

Figure 8.7 presents the rendering of a CFRP to produce a lyre in the CAD software Solidworks™. The interrelation of the design, the sound's quality, and the playability stand at the core of the musical instrument's development [20]. These factors along with the capabilities and the constraints of the manufacturing methods define the final form and performance of the product. During the last years, the use of CFRP to produce musical instruments has grown [21]. CFRP is stronger than steel but much lighter, while it is durable and steady in climatic changes, providing an uncompromised high-quality sound [22]. Furthermore, it sounds deeper and louder than traditional instruments [23]. The material is formed usually with the use of molds. The mold is treated with a layer of wax that allows the easy extraction of the product from the mold without leaving any residues [24]. Then, layers of carbon fiber and epoxy glue are applied. On the top of them a nylon woven fabric is laid and then a layer of a very porous material, which will absorb any extra epoxy glue, is placed. After that, the mold with the material is placed in a vacuum bag. The air is pumped out of the bag, and after the appropriate time the part is post-cured at 150°C. The excessive material is cut out with the use of diamond cutting wheels and the instrument is assembled, glued, finished, and calibrated. As it has been shown, carbon fiber materials provide advantages compared to wood in terms of musical instrument properties, processing technique, environmental sustainability, and sound quality. Carbon fiber instrument has great advance in parameters of mass, it can be more easily processed with lower

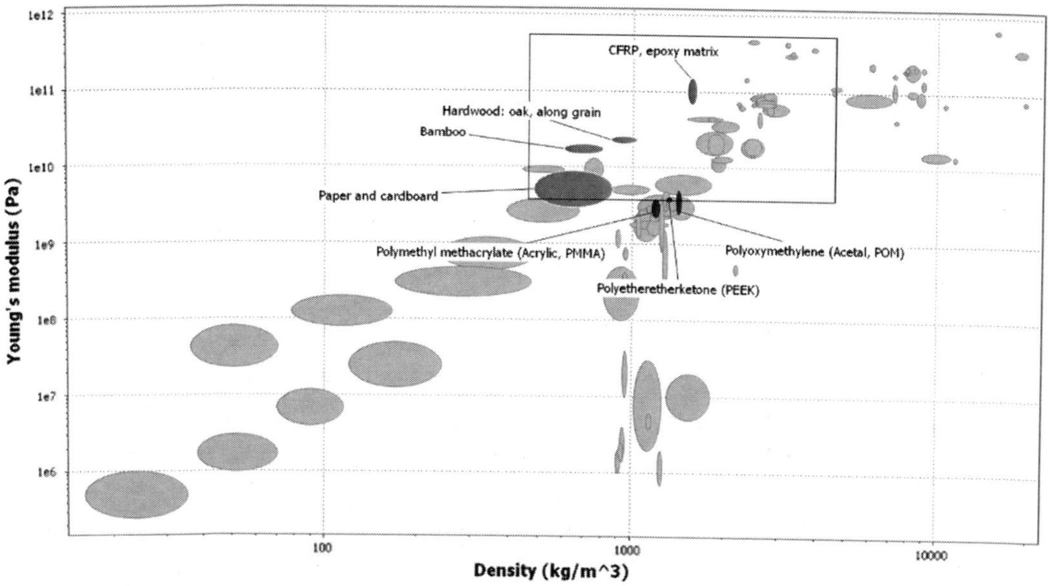

Figure 8.5 Rigid yet dense materials including woods.

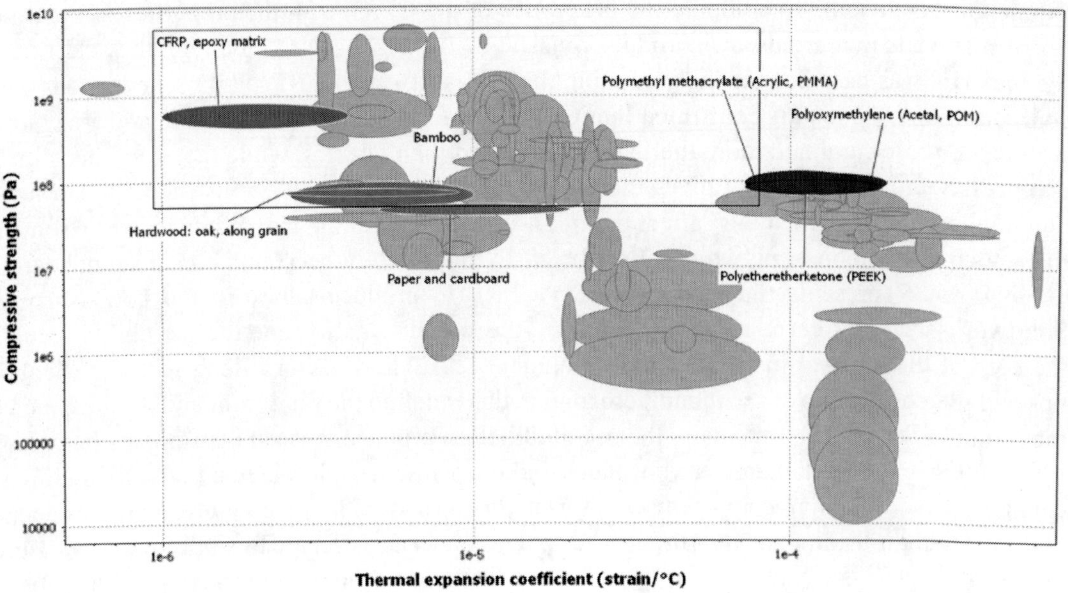

Figure 8.6 Compressively strong yet thermally unaffected materials including woods.

production time and cost, and it does not require the consumption of wood as well as quantities of colorants, paints, and glazes, and provides a more reach sound [25].

Vibromechanical Analysis of the New Pontic Lyre

The distinctive sound created by the lyre is a by-product of the vibrational input through the strings and the interaction of various parts of the lyre with each other. A finite element analysis (FEA) was conducted to examine the stress distributions and deflections on the main body of the lyre due to the forces applied by the strings, as well as to examine the vibrational characteristics. The materials that have been applied were several types of woods (oak for the sound box, fir for the soundboard, ebony for the bridge and the sound post) that are utilized in the traditional manufacturing of the pontic lyre, as well as CFRP, which has shown superior mechanical properties, as seen from the materials selection analysis. It is important to have a CFRP lay-up, designed such that out-of-plane deflection of the lyre would not be significant, in addition to match the modal frequencies of the wooden lyre. If the strings deflect the soundboard too much, then the result would be to make the strings go out of tune when fretted. Also, it is necessary to accommodate the stresses developed on the main body of the lyre through design modifications, as well as any undesired loadings that would interact with the vibrational behavior and lower the music quality of the instrument. The proposed novel lyre CAD model was analyzed in Solidworks™ simulation. Solidworks™ enables to apply boundary conditions and select various materials to simulate the structure and perform stress and vibrational analysis.

In general, the lyre's sound box has a rather complicated geometry due to its highly curved shape. For this reason, it was necessary to apply a mesh control to get more accurate results. Mesh control incorporates finer mesh that in turn results in more nodes and elements. For the static analysis, the regular mesh element size was 9.04161 mm whereas the mesh control

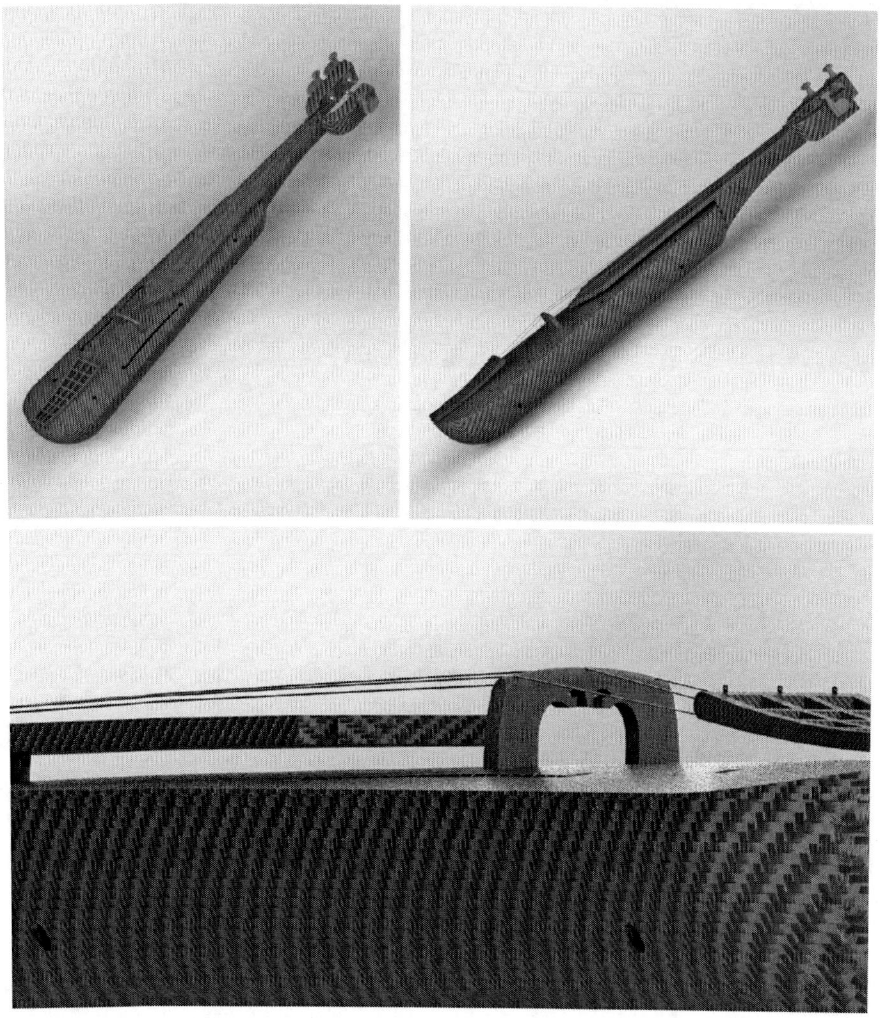

Figure 8.7 Rendering of a CFRP lyre.

applied created a finer mesh with an element size of 6.78121 mm. In total 18,409 elements and 35,865 nodes have been used in the current static study. For the vibrational simulation, both the classical (traditional) lyre and the new lyre have been analyzed. For the classical lyre regular mesh element size was 8.71042 mm whereas the mesh control applied created a finer mesh with an element size of 4.35521 mm. In total, 32,427 elements and 64,980 nodes have been used in the current vibrational study of the classical lyre. For the new lyre design, the regular mesh element size was 7.97087 mm, whereas the mesh control applied created a finer mesh with an element size of 3.98543 mm. In total, 32,421 elements and 64,771 nodes have been used.

For the static analysis, the top plate (soundboard) is bonded to the bottom one (sound box). Therefore, a fixed geometry was applied at the ends of the bottom plate. Additionally, three strings are rested on the bridge that transmits vibrations to the plates and are tied at both ends. A free-body diagram is sketched in Figure 8.8, illustrating the force distribution for a string on the bridge.

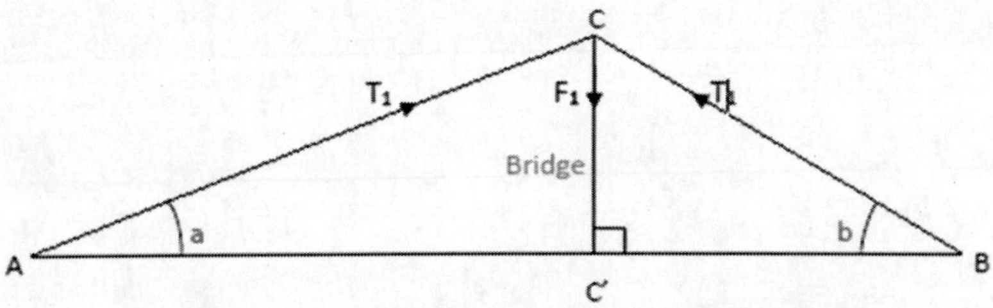

Figure 8.8 The free-body diagram of the forces applied from a string.

Table 8.1 The forces calculated from the analysis of the free-body diagram

1st string	2nd string	3rd string
A (La tone)	A (La tone)	B (Si tone)
Unit (N)	Unit (N)	Unit (N)
min–max values	min–max values	max value
48.89–63.51	48.89–63.51	61.1
F1 = 63.51*sin(1.06) + 63.51*sin(8.35) = 63.51*0.0185 + 63.51*0.145 = 1.174935 + 9.221652 = 10.396587	F2 = 63.51*sin(1.06) + 63.51*sin(8.35) = 63.51*0.0185 + 63.51*0.1452 = 1.174935 + 9.221652 = 10.396587	F3 = 61.10*sin(1.06) + 61.10*sin(8.35) = 61.10*0.0185 + 61.10*0.1452 = 1.13035 + 8.87172 = 10.00207

The load T_1 is the part of the excitation force due to the tension on the lyre of the string A (La tone). The string angles were calculated to be $a = 1.06°$ and $b = 8.35°$. Using the tension load T_1 and string angles, the static component of the force for string A (La tone) can be calculated using Equation (1):

$$F_1 = T_1 \sin\alpha + T_1 \sin b \tag{8.1}$$

Considering the loads of the three strings F_1, F_2, and F_3 the total force P acting on the lyre's bridge can be calculated from Equation (2):

$$P_1 = F_1 + F_2 + F_3 \tag{8.2}$$

Table 8.1 shows the results from the calculation of the forces. The appropriate string force was applied at the bridge as well as at the edges of the lyre (the latter only in the case of the static analysis). The total force of average lyre strings was $P = 30.8$ N.

The forces F_1, F_2, and F_3 were applied at each bridge position. Additionally, a total force of $P_2 = P_3 = 188.12$ N was applied on the top and bottom part of the lyre due to string tensioning for the static analysis.

Traditionally, the most common material for the manufacturing of the sound box is the plum (*Prunus domestica*). However, due to the lack of data regarding its mechanical properties (basically the material does not have any other commercial use), the properties of the oak wood, as it was mentioned above, were applied to the new model with the wooden materials. Oak wood was used since it is currently used for the manufacturing of lyres. Regarding the new model (wooden materials), the properties of the fir were used for the soundboard, those of the ebony for the bridge and the sound post, and the properties of the stainless steel (18/8 – AISI 304) were used for the anchor hook.

Carbon fiber varies greatly in its properties, depending on its manufacturing, the size of the bundles, and the type of weave. Carbon fibers generally have excellent tensile properties, low densities, high thermal and chemical stabilities in the absence of oxidizing agents, good thermal and electrical conductivities, and excellent creep resistance. They have been extensively used in composites in the form of woven textiles, prepregs, continuous fibers/rovings, and chopped fibers. The composite parts can be produced through filament winding, tape winding, pultrusion, compression molding, vacuum bagging, liquid molding, and injection molding [26]. It has been shown that with interlock and interweave bindings, CFRP component properties such as dimensional stability, compactness, and bending stiffness can be adjusted according to the requirements. Delamination at critical junctions can be prevented by engineered yarn paths of the weave and by crossing reinforcement threads between layers. The insertion of looping and crossing reinforcement threads, for example, closed reinforcement loops, allows an increase in mechanical properties of the CFRP component; especially, the delamination in critical areas of transverse loads can be reduced effectively [27].

The fabric that has been selected for the current analysis was a T300 plain weave which has a thickness of 0.25 mm. The thickness of the lyre and especially the sound box was 1 mm. Thereby, four layers of the T300 would suffice for the virtual construction of the lyre. In addition, the epoxy that was used was the Araldite LY5052 resin. Due to insufficient existing experimental material data, it was decided to use a laminate analysis software that is based on a classical lamination theory to predict the mechanical and vibrational properties of the CFRP materials and therefore receive reasonable results from the FEA. Three different fiber orientations were used in the laminate analysis software: $[0/90]_4$, $[\pm 45]_4$, and [0/90, ±45, 0/90, ±45]. The derived mechanical properties of the three laminates are shown in Table 8.2 along with the mechanical properties of the three woods (oak, fir, and ebony). All three CFRPs were applied to the new model in order to perform the tests.

Regarding the static analysis shown in Figure 8.9, the results have shown a hump on the top plate next to the bridge, caused by the forces acting from the strings via the bridge. This shape is usually visible to the naked eye on the classical wooden lyres as shown in Figure 8.10, which is attributed mainly to the induced creep from the local forces. In the depth of time, this phenomenon along with other unpredictable factors (humidity, normal wear, and others) gradually alters the qualitative sound characteristics of the instrument [10].

The maximum deflection was 0.1685040 mm for the new lyre design with wooden materials. Table 8.3 shows the maximum deflections at that local point obtained from the simulations. The stiffer CFRP materials have shown less maximum deflection at that local point. In fact, the higher Young's modulus, the smaller the deflection at that area. Soundboards of stringed instruments need to have a high Young's modulus/shear modulus ratio. That means a high elastic anisotropy contributes to sound radiation of the top plate because of a higher mobility of the vibration modes [28]. The low deflections indicate that the string force is not expected to deflect the lyre high enough to alter the playability of the instrument. From that point of view, the CFRP $[0/90]_4$ is the most suitable material for the manufacturing of the instrument. The entire lyre structure is anticipated to provide less stiffness compared with the rigid boundary conditions used in the simulation. Therefore, the absolute deflection of the top plate of the instrument, specifically the portion of bridge, which is not supported underneath by the sound post, is expected to be greater than what the numerical solution indicates.

While on the top of the lyre the deflection can be observed on the left side (Figure 8.9), on the bottom part, the deflection displacement appears in the area, where the sound post presses

Table 8.2 Material properties

Oak Quercus Robur (L)	
Yield strength	57,550 kN/m^2
Tensile strength	99,050 kN/m^2
Mass density	710 kg/m^3
Elastic modulus	13,750,000 kN/m^2
Poisson's ratio	0.375
Fir Abies Alba (L)	
Yield strength	39,900 kN/m^2
Tensile strength	84,000 kN/m^2
Mass density	445 kg/m^3
Elastic modulus	12,000,000 kN/m^2
Poisson's ratio	0.375
Ebony (L)	
Yield strength	28,900 kN/m^2
Tensile strength	127,500 kN/m^2
Mass density	1,040 kg/m^3
Elastic modulus	10,800,000 kN/m^2
Poisson's ratio	0.375
AISI 304	
Yield strength	206,810 kN/m^2
Tensile strength	517,020 kN/m^2
Mass density	8,000 kg/m^3
Elastic modulus	190,000,000 kN/m^2
Poisson's ratio	0.290
CFRP 0.90	
Yield strength	696,000 kN/m^2
Tensile strength	696,000 kN/m^2
Mass density	1,550 kg/m^3
Elastic modulus	58,460,000 kN/m^2
Poisson's ratio	0.300
CFRP ± 45	
Yield strength	159,000 kN/m^2
Tensile strength	159,000 kN/m^2
Mass density	1,550 kg/m^3
Elastic modulus	18,700,000 kN/m^2
Poisson's ratio	0.300
CFRP 0.90 ± 45	
Yield strength	312,000 kN/m^2
Tensile strength	312,000 kN/m^2
Mass density	1,550 kg/m^3
Elastic modulus	40,700,000 kN/m^2
Poisson's ratio	0.300

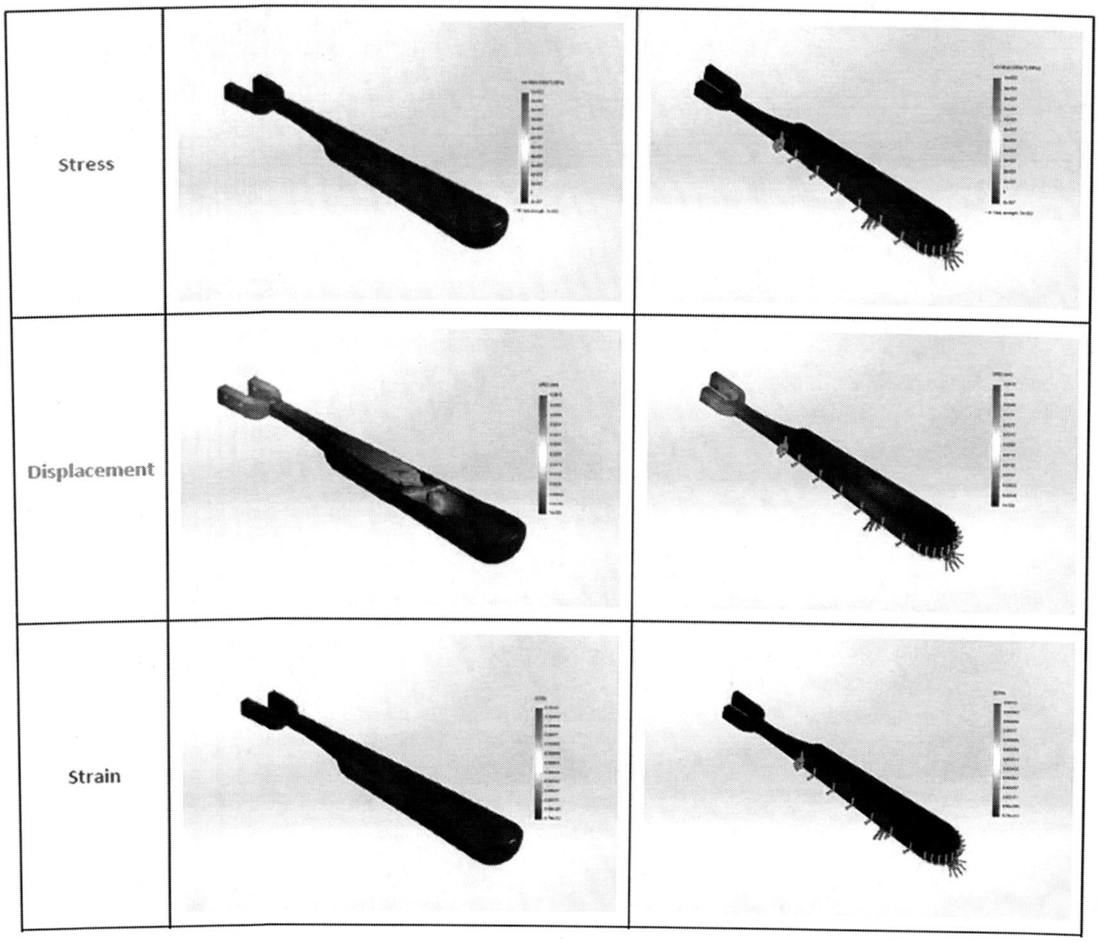

Figure 8.9 Stress, displacement, and strain of the model under the strings' forces.

(right side as seen from the top). This is an expected outcome, as the sound post's role is to transfer the energy from the top to the bottom of the instrument without causing undesirable and permanent deformation to the lyre.

Regarding the stresses, the analysis showed that the most critical area is the anchor hook of the chord holder (Figure 8.9). This is also expected, as all the tension of the strings is applied to that element. The results of Table 8.3 show that the AISI 304 stainless steel, which was applied to the new model with the wooden materials, is the strongest material for this use (maximum stress: 201,539,000 N/m^2). All three types of CFRP were quite close in terms of stress and they all have been within acceptable limits (no collapse of the product).

The examination of the vibrational behavior of an instrument is quite complicated [29]. In simple words, the strings vibrate the bridge and the soundboard, and vice versa (coupling) [30]. Accordingly, when the strings of the lyre are plucked, they start to vibrate at their natural frequency, and they transmit this vibration to the sound box through the bridge. The sound box transmits the vibration to the encapsulated air. The vibration is amplified and since the frequency of the vibration is audible, it is perceived as sound.

Regarding the vibrational behavior of the new lyre concept, if the natural frequency of the soundboard is the same as the forced vibration's frequency, the sound may become louder.

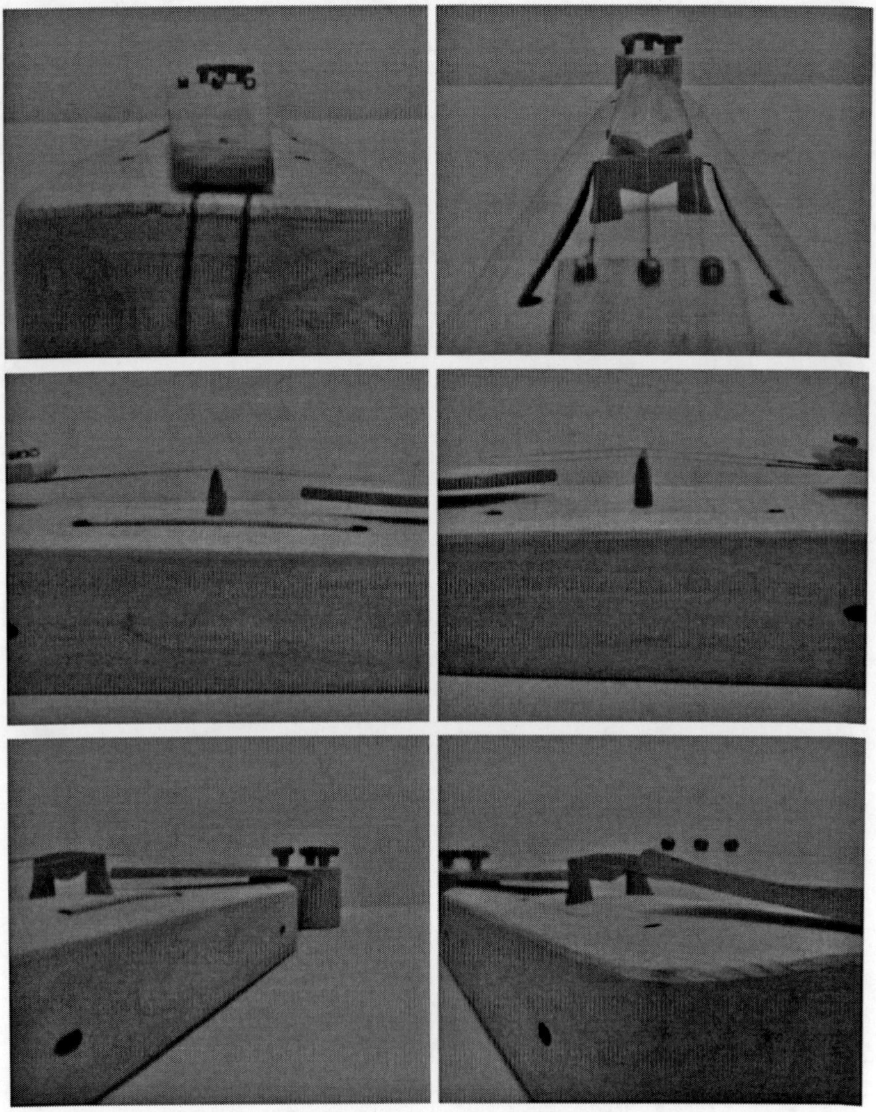

Figure 8.10 Induced creep (during 20 years) to the traditional lyre ("nostrils" area).

However, resonance effects with unexpected results may occur. For instance, the sound of a string may sound louder than all the other strings and this would make the instrument unplayable [31]. Additionally, low natural frequencies would not be acceptable because they result in decreased elastic moduli [32]. Furthermore, high frequencies depreciate quickly and are not contributing significantly to the sound's generation. Nevertheless, the sounds produced by the strings would be clearer.

A model of a classical sound box including a soundboard (lid) and a bridge was initially simulated. The resonance frequencies of this simulation are summarized in Table 8.4 while Figure 8.11 shows the mode shapes. The mode shapes show that the lyre's sides (ribs) have a limited vibration. The resonant frequencies are relatively low, and they are quite similar to the ones of a typical violin as was previously shown [33].

Table 8.3 Numerical results of the static analysis for the materials used

	Stress (N/m^2)	Displacement (mm)
New lyre with CFRP $[\pm45]_4$	98,589,000	0.1299310
New lyre with CFRP [0/90, ±45, 0/90, ±45]	98,630,200	0.0596677
New lyre with CFRP $[0/90]_4$	98,538,500	0.0415317
New wooden lyre	201,539,000	0.1685040

Table 8.4 Resonance frequency results for the various simulation models

Frequency number	Classical wooden lyre	New lyre with CFRP $[\pm45]_4$	New lyre with CFRP [0/90, ±45, 0/90, ±45]	New lyre with CFRP $[0/90]_4$	New wooden lyre
1	270.57	769.62	1,135.30	1,360.60	999.95
2	413.82	1,044.70	1,541.90	1,848.20	1,373.90
3	417.63	1,166.00	1,720.20	2,061.70	1,703.90
4	452.59	1,238.80	1,829.20	2,192.80	1,743.80
5	601.99	1,290.90	1904.80	2,282.90	1,810.30

In Figure 8.12 the mode shapes of the new lyre design are shown. The mode shapes are the same irrespective of the fiber orientation. The new lyre model with the CFRP material vibrates more extensively than the classical model. However, by running the analysis of the new model with the wooden materials (Figure 8.13), even though the three first modes have the same schematic behavior, differences occur in the last two modes. It is therefore revealed that the shape of the instrument plays a very significant role in its vibrating performance, as seen from the comparison of the classical and new lyre design, while modes 4 and 5 seem to be material-dependent.

Regarding the frequencies, the samples with the CFRP materials have higher values than that with the classical shape, as shown in Table 8.5. The frequencies of the new concept by applying the wooden materials are also significantly higher than that of the classical model and their values are comparable with those of the samples with CFRP fiber orientations $[\pm45]_4$ and [0/90, ±45, 0/90, ±45]. However, these materials have lower frequency modes than the CFRP $[0/90]_4$, which is the stiffest sample. The new design of the instrument enables higher resonant frequencies than the classical one. The frequencies of a structure are proportional to its stiffness and inversely proportional to its mass. Therefore, the higher resonant frequencies are attributed to the higher stiffness of the CFRP materials and the lower mass of the new lyre design. More specifically the $[0/90]_4$ fiber orientation has the highest Young's modulus enabling the highest resonant frequencies. Additionally, the mass of the new lyre design is 96 gr, compared with the 106.4 gr of the classical lyre.

In view of the above, in order to achieve high resonance frequencies enabling the generation of clearer-brighter tones, the sound post of the lyre was changed (Figure 8.14), while all the other parameters of the simulation such as the material and fixtures loads were kept constant. By applying the finest mesh, the frequencies were slightly increased as shown in Table 8.5, while the vibrating schema changed, for the last three modes as shown in Figure 8.15.

Mode Shape: 1 (Value= 270.568 Hz)	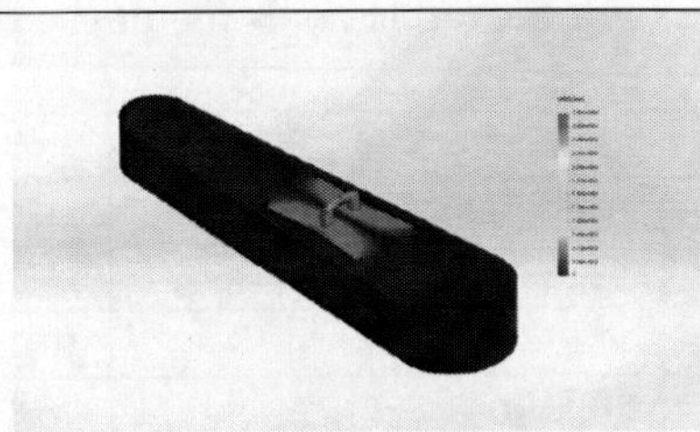	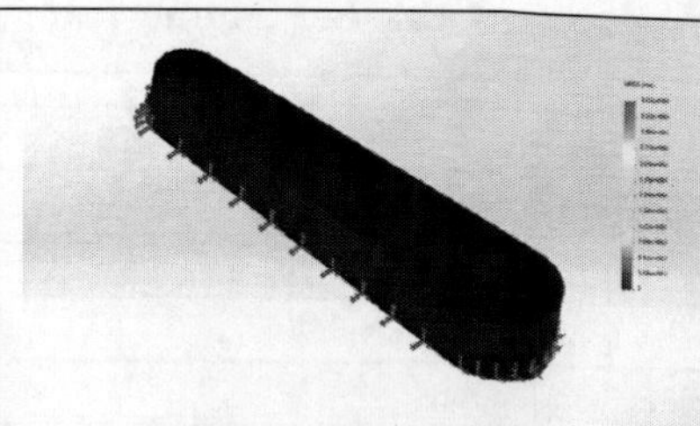
Mode Shape: 2 (Value = 413.821 Hz)	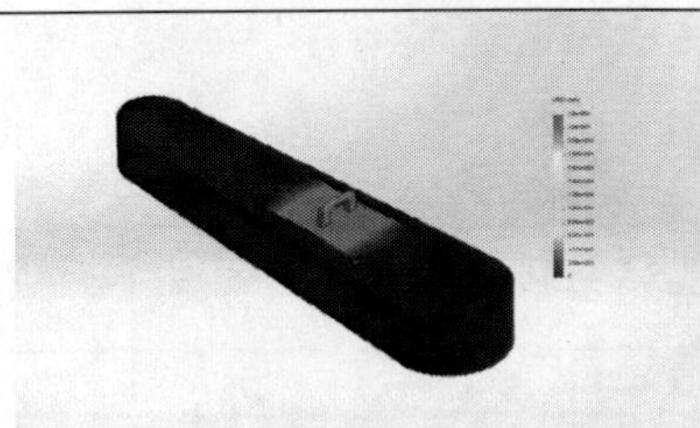	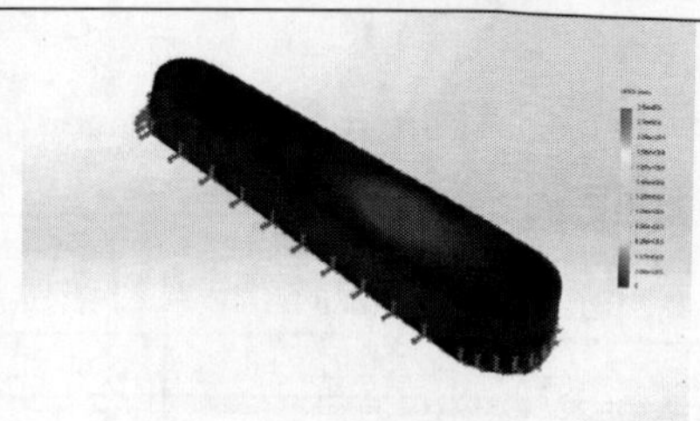
Mode Shape: 3 (Value = 417.628 Hz)		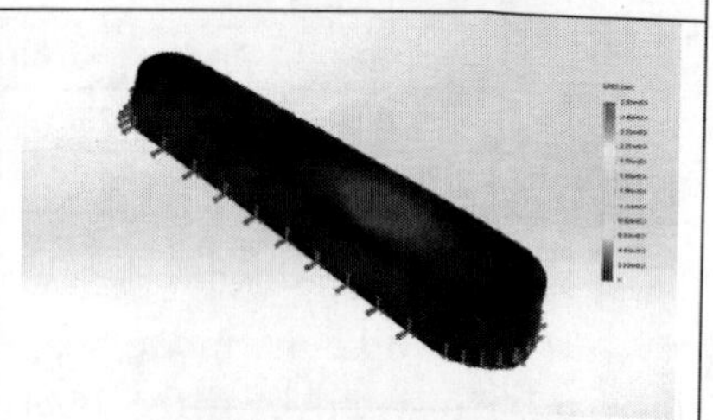
Mode Shape: 4 (Value = 452.59 Hz)	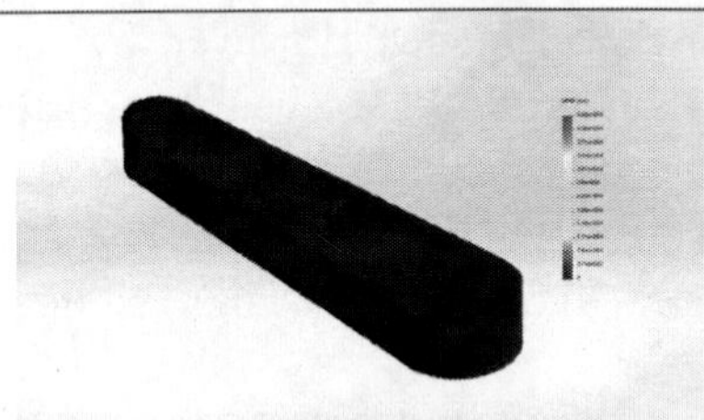	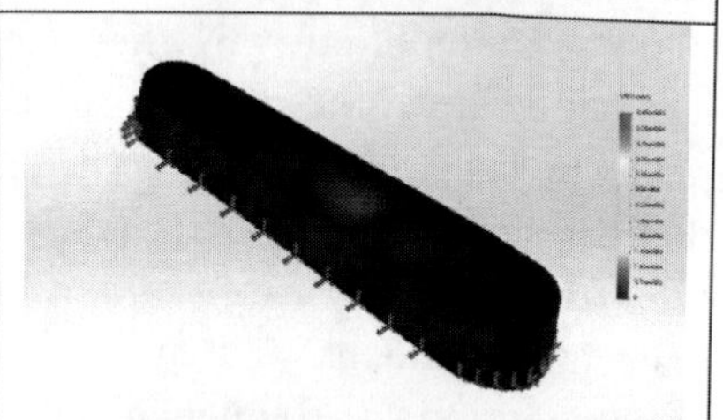
Mode Shape: 5 (Value = 601.988 Hz)	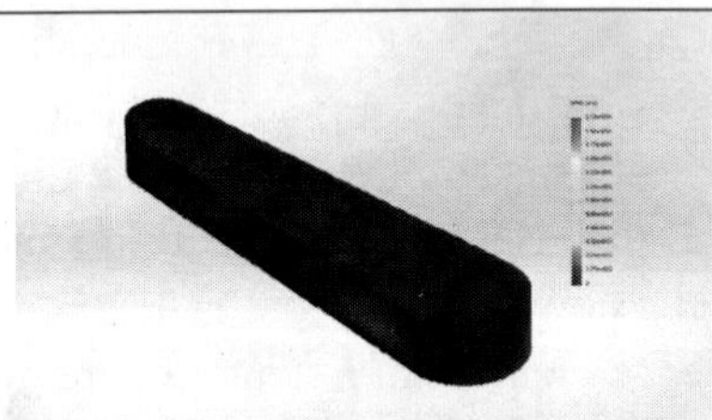	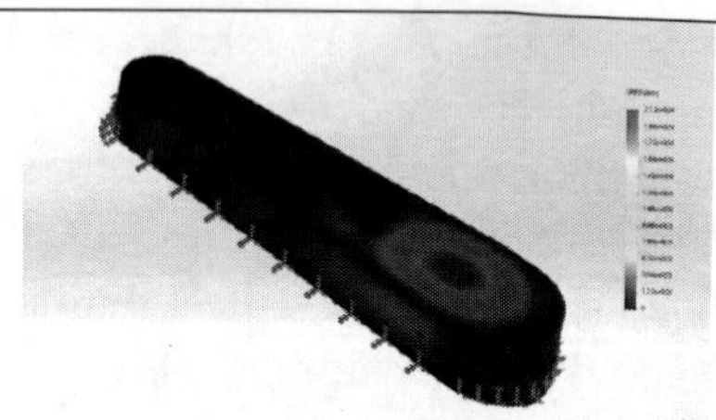

Figure 8.11 The mode shapes 1–5 of the classical lyre.

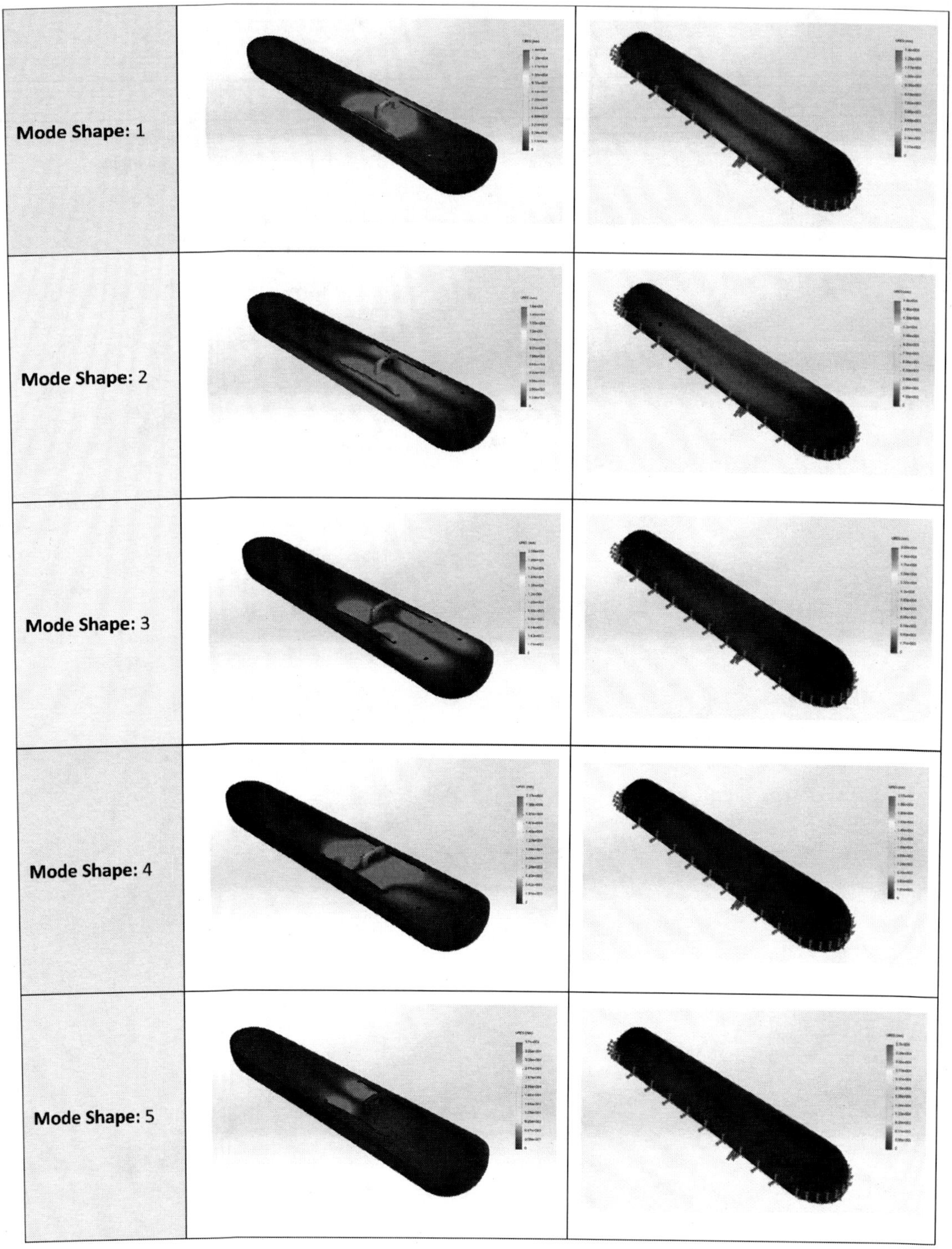

Figure 8.12 The mode shapes 1–5 of the new lyre design with CFRP.

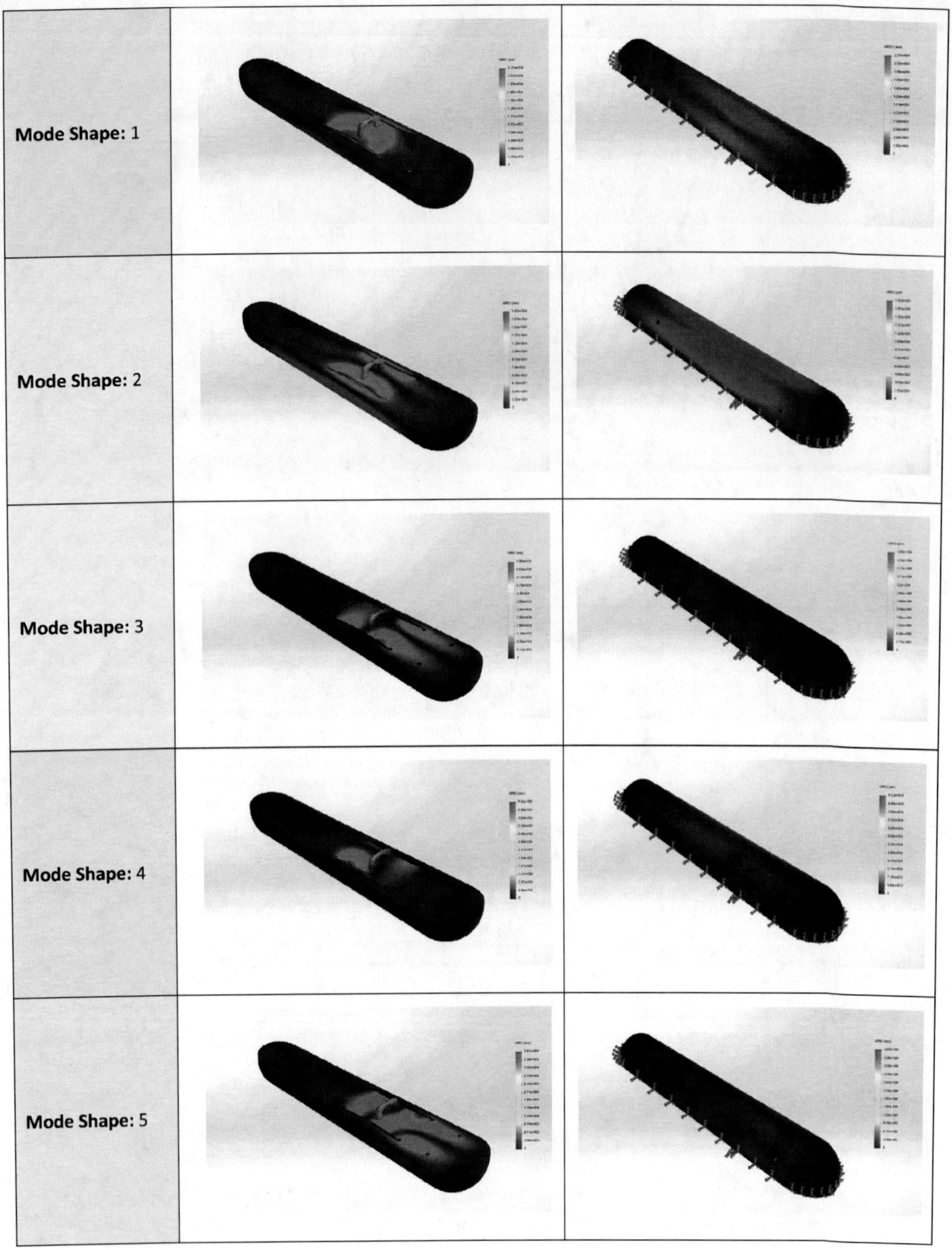

Figure 8.13 The mode shapes 1–5 of the new lyre design with wood.

Table 8.5 Resonance frequency results for the new model with the extended bridge

Frequency number	New extended bridge lyre with CFRP [±45]$_4$	New extended bridge lyre with CFRP [0/90, ±45, 0/90, ±45]	New extended bridge lyre with CFRP [0/90]$_4$
1	782.05	1,153.80	1,382.90
2	1,075.70	1,588.00	1,903.50
3	1,266.90	1,870.60	2,242.40
4	1,358.80	2,005.80	2,404.30
5	1,384.30	2,044.10	2,450.40

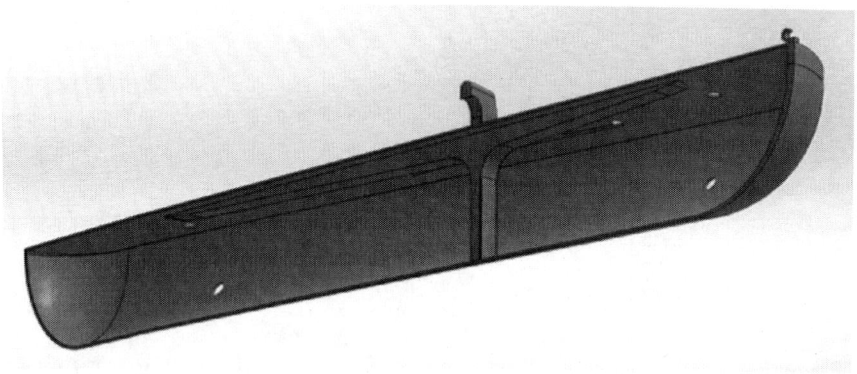

Figure 8.14 The new lyre with the extended bridge.

Conclusion

Regarding the static analysis, the results have shown a hump on the top plate next to the bridge, caused by the forces acting from the strings via the bridge. This is attributed mainly to the induced creep from the local forces. The stiffer CFRP materials have shown less maximum deflection at that local point. The higher Young's modulus, the smaller the deflection at that area. The low deflections indicate that the string force is not expected to deflect the lyre high enough to alter the playability of the instrument. From that point of view, the CFRP [0/90]$_4$ is the most suitable material for the manufacturing of the instrument. On the bottom part, the deflection displacement appears in the area where the sound post presses as it transfers the energy from the top to the bottom of the instrument without causing undesirable and permanent deformation to the lyre. Furthermore, in terms of the stresses, the analysis showed that the most critical area is the anchor hook of the chord holder as all the tension of the strings is applied to that element. Regarding the vibrational behavior of the new lyre concept, the mode shapes showed that the lyre's sides (ribs) have a limited vibration. The use of CFRP materials leads to more extensive vibrations and it is revealed that the shape of the instrument plays a very significant role in its vibrating performance. The CFRP sample with fiber orientation [0/90]$_4$ is the stiffest one and exhibits the highest frequency modes. The redesign of a traditional musical instrument is very challenging and intriguing. Many research works focus on the use of new materials without altering the form of the instrument. In this study, we tried to go a step further optimizing the form of the instrument to improve its performance. During the redesigning process, many morphological possibilities and alternatives can be used to keep the balance between

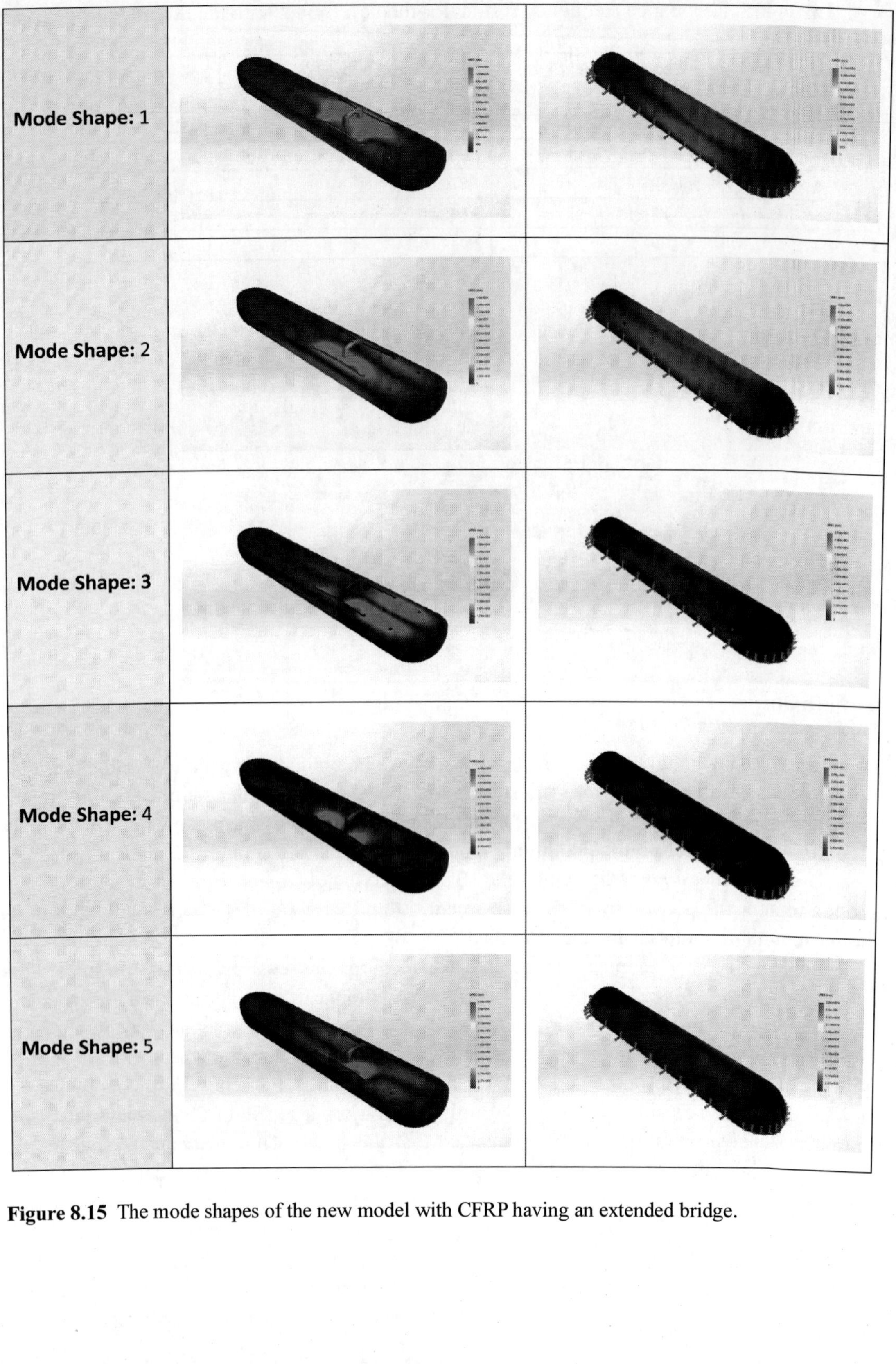

Figure 8.15 The mode shapes of the new model with CFRP having an extended bridge.

novelty and tradition. Furthermore, the material selection process offers several options with the final choice depending on several factors such as the cost-efficiency of the materials processing and machining. The current study provided a solid methodology for the development of new age lyre including the design, the material selection, and the simulation processing. It provided a theoretical model for further analysis and implementation, and it can be considered as a road-map for similar projects targeting testing, prototyping, and standardizing of the new age lyre.

References

[1] J. Lienhard, "Music for a while," 2004. [Online]. Available: https://uh.edu/engines/musicforawhile/musictext.htm.

[2] parodos.net.gr, "epapado2002<at>sch.gr," 2014. [Online]. Available: https://parodos.net.gr/elepapado/images/Instruments/Pandouris.jpg.

[3] M. Kaliontzidis, *The Lyre of Pontus. Instrumentation, Technique, Repertoire*, Athens: Fagotto, 2008.

[4] M. Tsahouridis, "The Pontic lyra in contemporary Greece," PhD Thesis, London: Goldsmiths University of London, 2007.

[5] I. Beina, "The lyre of Karpathos: An ethnographic approach to cultural instrumentation. Musical instruments and cultural identity of Karpathos," PhD Thesis, Ionian University, Corfu, Greece, 2011.

[6] paradosiakaorgana.gr, "paradosiakaorgana.gr," 2014. [Online].

[7] Pontos News, "Pontic lyre," 2014. [Online]. Available: http://www.pontos-news.gr/pontic-article/100058/pontiaki-lyra.

[8] J. Alexandris, "Lyras," 2009. [Online]. Available: https://sites.google.com/site/alexandrisjannis/Home/ouds/greek-traditional-lute/lavdas/lyras.

[9] K. R. Zappas, "The science of sound: Examining the role of materials in musical instruments," *JOM,* 59, pp. 13–17, 2007.

[10] V. Bucur, *Handbook of Materials for String Musical Instruments*, Switzerland: Springer, 2016.

[11] C. Y. Barlow, "Materials selection for musical instruments," *Musical Acoustics: ISMA'97,* 5, pp. 69–78, 1997.

[12] S. S. J. Kopac, "Wood: an important material in manufacturing technology," *Journal of Materials Processing Technology*, 133, pp. 134–142, 2003.

[13] D. E. Kretschmann, "Mechanical properties of wood," in *Wood Handbook - Wood as an Engineering Material*, Centennial Edition ed., Madison, Forest Products Laboratory, 2010, pp. 100–145.

[14] K. Sardis and M. Papachristopoulou, *Acoustical Properties of Wood-Manufacturing of a String Musical Instrument*, University of Thessaly, Karditsa, Greece, 2005.

[15] I. Xanthopoulos, "How to create a pontic lyre," 1993. [Online]. Available: https://www.youtube.com/watch?v=MBV4SZcVxsw.

[16] Greek Luthiers Community, "Greek luthiers community," 2013. [Online]. Available: http://www.luthier.gr/index.php?topic=1251.msg19698#msg19698.

[17] V. Vasileiadis, "Vasileiadis guitars," [Online]. Available: http://www.vasileiadisguitars.com/projects/kemence_construction.pdf. [Accessed July 14, 2014].

[18] M. Kaliontzidis, "Pontic lyre," 2012. [Online]. Available: http://www.pontiakilyra.gr/2012/02/blog-post_02.html. [Accessed June 6, 2020].

[19] W. D. Callister Jr. and D. G. Rethwisch, *Materials Science and Engineering. An Introduction*, 8th ed., Hoboken, New Jersey: John Wiley & Sons, Inc., 2010.

[20] A. Zoran, "The 3D printed flute: Digital fabrication and design of musical instruments," *Journal of New Music Research*, 40(4), pp. 379–387, 2011.

[21] L. D.-B. a. F. S.-L. Manuel Ibáñez-Arnal, "A study of the dynamic response of Carbon Fiber Reinforced Epoxy (CFRE) prepregs for musical instrument manufacturing," *Applied Sciences*, 9, p. 4615, 2019.

[22] "Luis and Clark. Carbon fiber instruments," [Online]. Available: https://luisandclark.com/the-instruments/. [Accessed June 7, 2020].

[23] H. K. Scott, "Strings magazine," 2009. [Online]. Available: http://stringsmagazine.com/a-buyers-guide-to-carbon-fiber-instruments/. [Accessed June 7, 2020].

[24] M. S. Probert, *Design, Manufacture and Analysis of a Carbon Fiber Epoxy Composite Acoustic Guitar*, Seattle, WA: University of Washington, 2007.

[25] J. H. L. Ze Hong Wu, "Carbon fiber material in musical instrument making," *Materials and Design*, 89, pp. 660-664, 2016.

[26] X. Huang, "Fabrication and properties of carbon fibers," *Materials,* 2, 2009, 2369–2403.

[27] H.-J. B. J. W. J. S. N. S. L. v. W. M. T. a. G. T. G. Stefan Schindler, "Engineering of three-dimensional near-net-shape weave structures for high technical performance in carbon fibre–reinforced plastics," *Journal of Engineered Fibers and Fabrics*, 14, pp. 1-16, 2019.

[28] M. Z. A. W. Robert Sproßmann, "Characterization of acoustic and mechanical properties of common tropical woods used in classical guitars," *Results in Physics*, 7, pp. 1737-1742, 2017.

[29] J. W. C. H. Hodges, "Theories of noise and vibration transmission in complex structures," *Reports on Progress in Physics*, 49, p. 107, 1999.

[30] N. H. F. T. D. Rossing, *Principles of Vibrationand Sound*, New York: Springer, 2004.

[31] R. Bader, *Computational Mechanics of the Classical Guitar*, New York: Springer, 2005.

[32] P. P. L. G. E. M. A. Perez Martinez, "Vibration testing for the evaluation of the effects of moisture content on the in-plane elastic constants of wood used in musical instruments," in *Vibration and Structural Acoustics Analysis*, D. R. C. M. A. Vasques, Ed., New York, Springer, 2011.

[33] T. Hong, *A Vibration Study of Violins*, Singapore, National University of Singapore, 2011.

About the Editors

Dr. Panagiotis Kyratsis (www.kyratsis.com) is Professor in the Department of Product and Systems Design Engineering, University of Western Macedonia, Greece. He is Director of the Institute of Traditional Architecture and Cultural Heritage, University Research Center "TEMENUS," and Director of the Computational Design and Digital Fabrication Research Lab. Dr. Panagiotis Kyratsis received his Ph.D. in the area of CAD-based manufacturing process simulations from the School of Production Engineering and Management, Technical University of Crete, Greece. He holds a diploma in Mechanical Engineering from Aristotle's University of Thessaloniki, Greece, and he received his M.Sc. in Automotive Product Engineering and M.Sc. in CAD/CAM from Cranfield University, UK, in 1997 and 1999, respectively. He has involved in a number of industrial projects, and he has a great deal of expertise in both the design and manufacturing aspects of product development. His main research interests include manufacturing, machining, CAD/CAM/CAE systems, product design, reverse engineering, and prototyping. He has published more than 20 books and more than 160 papers in scientific journals and international conferences. He acts as a member of the editorial board and reviewer to numerous scientific journals and holds twelve industrial designs and one patent registered within the Greek Patent Office.

Dr. Dimitrios Tzetzis is Associate Professor in the Department of Science and Technology, International Hellenic University, Greece. He is Director of the Digital Manufacturing and Materials Characterization Laboratory and Director of the M.Sc. in Strategic Product Design postgraduate programme. Dr. Dimitrios Tzetzis graduated from the Department of Mechanical Engineering, Kingston University, London, UK, and he continued his studies at the same university, receiving an M.Sc. in Advanced Manufacturing Systems. He also received his Ph.D. from Queen Mary, University of London, School of Engineering and Materials Science, UK, where he focused his research on the design, analysis, and manufacture of materials and structures. Currently, apart from his position at International Hellenic University, he is Visiting Professor in various other educational establishments. He participates in national- and international-funded research projects in the field of mechanical analysis of advanced materials and digital manufacturing methods for various applications. He is a member of scientific committees at the national and international levels, he acts as a member of the editorial board and reviewer to numerous scientific journals, and he is the author of more than 120 scientific publications in peer-reviewed journals, 9 chapters in edited books, and more than 100 publications in international conferences.

Index

D

E

F

G

H

S

T

U

V

W

Z